PLANT SCIENCE RESEARCH AND PRACTICES

PHASEOLUS LUNATUS

DIVERSITY, GROWTH AND PRODUCTION

PLANT SCIENCE RESEARCH AND PRACTICES

Additional books in this series can be found on Nova's website under the Series tab.

Additional e-books in this series can be found on Nova's website under the e-book tab.

Plant Science Research and Practices

Phaseolus Lunatus

Diversity, Growth and Production

Angela Celis de Almeida Lopes
Regina Lucia Ferreira Gomes
and
Ademir Sergio Ferreira de Araujo
Editors

nova publishers
New York

Library of Congress Cataloging-in-Publication Data

Phaseolus lunatus : diversity, growth and production / editors: Ademir Sirgio Ferreira de Arazjo, Angela Celis de Almeida Lopes, and Regina Lucia Ferreira Gomes, Department of Agricultural Engineering and Soil Science, Agricultural Science Center, Federal University of Piaum Teresina, PI, Brazil.
pages cm. -- (Plant science research and practices)
Includes index.
ISBN 978-1-63482-494-1 (hardcover)
1. Lima bean. I. Arazjo, Ademir Sirgio Ferreira de, editor. II. Series: Plant science research and practices.
SB327.P43 2015
635'.653--dc23
2015009604

Published by Nova Science Publishers, Inc. † New York

CONTENTS

PREFACE

Of the 200,000 to 300,000 plant species, less than 1,000 have been domesticated, if only partially. Most of the research and funding attention goes to a few major crop plants that are grown over large areas (often most continents) and play an important role in human sustenance, such as the three major cereals: maize, rice, and wheat. Yet, it is well known that nearly exclusive reliance on a limited number of crops can have negative consequences from an agronomic and human nutritional standpoint. Crop biodiversity provides food security[1], pest and disease management, improved soil processes, increased food for pollinators, reduced soil erosion, increased nitrogen fixation, improved soil sequestrations, and resiliency when faced with major disturbances.[2,3] Agricultural biodiversity also promotes the chances of a diverse diet and, in turn, may improve the health status of humans, especially of children[4]. Furthermore, the global climate change to which agriculture is currently subjected to, especially at low latitudes, calls for exploration of other crops that may be better adapted to radically changed growing conditions such as an increase in temperature of 3-4°C and increased rainfall uncertainty[5] and can contribute to more sustainable dietary changes[6]. One of these crops, lima bean (*Phaseolus lunatus*) is the second most important domesticated species in the genus *Phaseolus* after common bean (*P. vulgaris*). It is grown principally as a subsistence crop in hotter climates of the tropics, such as in Lowland

---

[1] Esquinas-Alcazar J (2005) Nature Reviews Genetics 6:946-953

[2] Hajjar R, Jarvis DI, Gemmill-Herren B (2008) Agriculture, Ecosystems & Environment 123:261-270

[3] Warschefsky E, Penmetsa RV, Cook DR, von Wettberg EJB (2014) American Journal of Botany 110 (10): 1791-1800

[4] Frison EA, Cherfas J, Hodgkin T (2011) Sustainability 3:238-253

[5] Rosenzweig C et al. (2014) Proceedings of the National Academy of Sciences 111:3268-3273.

[6] Tilman D, Clark M (2014) Nature 515:518-522

South America and Africa (in addition to localized commercial production in other regions, such as California and Delaware in the USA). This volume is devoted entirely to current research on lima bean by experts in the field. It consists of nine chapters on a broad range of topics relevant to improved production, including the native diversity of lima beans in Brazil and Mexico, vegetative and reproductive development, fertilization, symbiotic nitrogen fixation, diseases, and socio-economic importance. Let us hope that this volume is the prelude to an international collaborative network to advance lima bean production, consumption, and research as a response to global change!

Paul Gepts
University of California, Davis

List of Contributors

Jaime Martínez Castillo, PhD
Centro de Investigación Científica de Yucatán, A. C. Calle 43 No. 130, Col. Chuburna de Hidalgo, Mérida, Yucatán 97200, México.

Rubén Humberto Andueza Noh, PhD
División de Estudios de Posgrado e Investigación, Instituto Tecnológico de Conkal-CATEDRAS-CONACYT. Avenida Tecnológico, Conkal, Yucatán 97345.

María Isabel Chacón Sánchez, PhD
Facultad de Ciencias Agrarias, Universidad Nacional de Colombia. Carrera 30 No. 45-03, Edificio 500, Bogotá, D.C., Colombia.

Maria da Conceição Prado de Oliveira, PhD
Departamento de Biologia, Universidade Federal do Piauí, Teresina, Brazil.

Ana Clara de Sousa Meirelles, PhD
Department of Agronomy, Universidade Estadual de Maringá, Maringá, Brazil.

Ângela Celis de Almeida Lopes, PhD
Departamento de Biologia, Universidade Federal do Piauí, Teresina, Brazil.

Sergio Emilio Valente
Federal University of Piauí, Centre of Agricultural Sciences, Department of Plant Science, Teresina, Piauí, Brazil

Lidiane Lima Feitoza
Federal University of Piauí, PNPD/CAPES – Pos Graduate Program in Genetics and Improvement, Teresina, Piauí, Brazil

Marcones Ferreira Costa, MsC
Programa de Pós Graduação em Genética e Melhoramento, Universidade Federal do Piauí, Teresina, Brazil.

Ademir Sérgio Ferreira de Araújo, PhD
Agricultural Science Center, Federal University of Piauí, Teresina, Piauí, Brazil

Sandra Mara Barbosa Rocha, BsC
Agricultural Science Center, Federal University of Piauí, Teresina, Piauí, Brazil

Dyego Leandro da Costa Monte, BsC
Agricultural Science Center, Federal University of Piauí, Teresina, Piauí, Brazil

Nilza Silva Carvalho, MsC
Agricultural Science Center, Federal University of Piauí, Teresina, Piauí, Brazil

Ernesto Ormeño-Orrillo, PhD
Center for Genomic Sciences, Universidad Nacional Autónoma de México, Cuernavaca, Morelos, Mexico.

Doris Zúñiga Dávila, PhD
Microbial Ecology and Biotechnology Laboratory, Department of Biology, Universidad Nacional Agraria La Molina, Lima, Peru.

Esperanza Martínez-Romero, PhD
Center for Genomic Sciences, Universidad Nacional Autónoma de México, Cuernavaca, Morelos, Mexico.

Jadson Emanuel Lopes Antunes, PhD
Soil Biology Laboratory, Agronomical Institute of Pernambuco, Recife, Pernambuco, Brazil

Regina Lucia Ferreira Gomes, PhD
Agricultural Science Center, Federal University of Piauí, Teresina, Piauí, Brazil

Marcia Barreto do Vale Figueiredo, PhD
Soil Biology Laboratory, Agronomical Institute of Pernambuco, Recife, Pernambuco, Brazil

Eulalia Maria de Sousa Carvalho, PhD
Agricultural Science Center, Federal University of Piauí, Teresina, Piauí, Brazil

José Evando Aguiar Bezerra, PhD
Agricultural Science Center, Federal University of Piauí, Teresina, Piauí, Brazil

Beatriz Meirelles Barguil, PhD
State University of Piauí, Teresina, Piauí, Brazil

Jardel Oliveira Santos, PhD
Federal University of Maranhão, Chapadinha, Maranhão, Brasil

Cristiane Araújo Soares, MsC
State University of São Paulo, São Paulo, Brazil

Francisco Carlos Gândara
Agricultural Science Center, Federal University of Piauí, Teresina, Piauí, Brazil

Marília Lobo Burle
EMBRAPA Cenargen, Brasilia, DF Brazil

In: Phaseolus Lunatus
Editors: A. S. Ferreira de Araujo et al.
ISBN: 978-1-63482-494-1

Chapter 1

GENETIC DIVERSITY OF LIMA BEAN

***Ângela Celis de Almeida Lopes*[*1], *Regina Lucia Ferreira Gomes*[2], *Sérgio Emílio dos Santos Valente*[1] *and Lidiane de Lima Feitoza*[3]**

[1]Federal University of Piauí, Centre for Natural Sciences, Department of Biology, Teresina, Piauí, Brazil
[2]Federal University of Piauí, Centre of Agricultural Sciences, Department of Plant Science, Teresina, Piauí, Brazil
[3]Federal University of Piauí, PNPD/CAPES – Pos Graduate Program in Genetics and Improvement, Teresina, Piauí, Brazil

ABSTRACT

The maintenance of genetic variability in populations is the basis of the conservation of species; therefore, its description and distribution are crucial for the establishment of truly effective conservation practices. Because this variability is observed in organisms and transmitted through reproduction via genetic recombination, each individual can be considered to be apparently unique in genetic constitution relative to their own species and much more unique relative to other species. Brazil is

[*] Corresponding author address: Dr. Angela de Almeida Lopes. Nature Science Center, Federal University of Piauí, Teresina, PI (Brazil); Tel.: +55-86-3215-5700. E-mail address: acalopes@ufpi.edu.br.

among the countries with the greatest genetic variability available for agricultural and related research, but the use of this variability is incipient in most cases, and various tools are required to characterise this variability.

In this chapter will be presented information about characterization and study of genetic diversity in lima bean. The data allowed to observe that the morphological markers can provide adequate characterization of the available variability and permit a greater use of genotypes maintained in the active germplasm banks (AGBs).

INTRODUCTION

One of the goals of genetic studies of populations is to describe the amount of genetic variation within a population. This variation is fundamental for evolution because natural selection acts on variants that occur within a population as a result of adaption to the environment; these variants produce variation among populations and ultimately variation between species [17]. Thus, greater genetic variability in a population increases the likelihood of the populations' perpetuation. The maintenance of genetic variability in populations is the basis of the conservation of species [20]; therefore, its description and distribution are crucial for the establishment of truly effective conservation practices.

Genetic variation can be characterised and identified based on phenotypic factors, such as morphological traits, DNA and its fragments, mutant genes, chromosomes, genetic markers, such as isoenzymes, PCR, Random Amplified Polymorphic DNA (RAPD), Restriction Fragment Length Polymorphism (RFLP), Amplified Fragment Length Polymorphism (AFLP), Sequenced Characterised Amplified Region (SCAR), Sequence Tagged Sites (STS) and microsatellites [7]. Because this variability is observed in organisms and transmitted through reproduction via genetic recombination, each individual can be considered to be apparently unique in genetic constitution relative to their own species and much more unique relative to other species. Thus, there may be individuals with genotypes more adapted to certain conditions.

A study of the genetic variability of lima bean using 22 isozyme loci in 29 wild populations from the Central Valley of Costa Rica showed that the proportion of polymorphic loci (P_p), mean number of alleles per locus (A) and mean number of effective alleles per locus (A_e) were equal to 10.32%, 1.10 and 1.05, respectively [22]. The heterozygosity (H_T) was 0.193 and intrapopulation (H_S) and interpopulation (D_{ST}) genetic diversity were 0.082

and 0.111, respectively. In that study, three specific alleles were identified in different populations, which emphasises the importance of diversity studies. The genotypic composition of the population was analysed, and it exhibited deviations from Hardy-Weinberg equilibrium (F_{IT} = 0.932) because of genetic differentiation among populations (F_{ST} = 0.497) or non-random mating within populations (F_{IS} = 0.866), which indicates that these populations are autogamous.

In predominantly autogamous and short-lived perennial species, such as wild lima bean, gene differentiation between populations (G_{ST}) is typically high [8]. The wild lima bean has a mixed mating system and is a predominantly self-pollinated species. Isoenzyme electrophoresis data from wild lima bean populations have indicated that genetic variability primarily occurs at the interpopulation level, with low observed and expected heterozygosity and low interpopulation gene flow [21].

Brazil is among the countries with the greatest genetic variability available for agricultural and related research, but the use of this variability is incipient in most cases, and various tools are required to characterise this variability. Morphological and molecular markers can provide adequate characterisations of the available variability and permit a greater use of the accessions maintained in the active germplasm banks (AGBs).

Morphological Characterization and Study of Genetic Diversity in Lima Bean Accessions

Morphological characterisations can identity genotypes based on data sets that provide the genetic variability of each sample, and such characterisations are required for the effective use of genetic resources conserved in germplasm banks; according to Valls [18], more than 200,000 accessions of various plant species are stored in banks scattered across the country, and the minimum information necessary for their use in breeding programs is not available.

The morphological study of lima bean varieties is important because it can identify diagnostic traits and provide access to this material in searches of plants that have adequate productivity and performance in different environmental conditions [14].

In a study conducted in the 2005/2006 agricultural year, 85 lima bean accessions from the Active Lima Bean Germplasm Bank, Federal University of Piauí (Banco Ativo de Germoplasma de Feijão-Fava da Universidade Federal do Piauí - UFPI-BAGF) were multiplied in the experimental area of the Department of Plant Science, UFPI Agricultural Science Centre, in Teresina City, Piauí (PI), which is located at 05°05'05" S latitude, 42°05' W longitude at an altitude of 72.7 m. Four seeds were sown per hole, and a single plant was maintained after thinning and supported with a wooden stake. The plots consisted of five-metre rows with 1.0 m x 1.0 m spacing. At this stage, the accessions were characterised according to the following descriptors published by the International Plant Genetic Resources Institute [10], currently Biodiversity International: pod pubescence, pod beak shape, pod length, pod width, pod curvature, pod colour, seeds per pod, loci per pod, and 100 seed weight.

Among the evaluated descriptors, 52.32% of the genotypes had a thick pod beak and 47.68% had a short pod beak. The length of the pod ranged from 4.07 cm to 9.20 cm, with a mean and variance of 6.46 cm and 1.64 cm, respectively. The pod width had minimum, maximum and mean values of 1.03 cm, 1.90 cm and 1.46 cm, respectively, with a variance of 0.04 cm. The seeds per pod showed minimum, maximum, mean and variance values of 1.67, 3.00, 2.38 and 0.21, respectively (Table 1). The observed levels of variance have the potential for use in the selection of genotypes with a higher seeds per pod. Lemos et al. [11] found that common bean seed yield was correlated with the seeds per pod, which is an important component of bean production.

The 100 seed weight ranged from 24.95 g to 88.66 g, averaged 47.65 and had a variance of 341.65. According to Vieira [19], the 100 seed weight can range from 30 to 300 g, and this large variance is favourable for the selection of genotypes with different seed sizes and allow the breeder to develop cultivars that meet the variable preferences of the consumer market, which was observed in the testimonies of merchants interviewed during collection trips held in various municipalities of Piauí.

Table 1. Mean, minimum, maximum and variance values of the phenotypic traits pod length, pod width, seeds per pod and 100 seed weight assessed in 85 lima bean accessions (Teresina, PI, 2005/2006)

Trait	Mean	Minimum	Maximum	Variance
Pod length (cm)	6.47	4.30	9.30	1.64
Pod width (cm)	1.47	1.09	2.00	0.04
Seeds per pod	2.43	1.50	4.00	0.21
100 seed weight (g)	47.65	24.95	88.66	341.65

Data from 44 of the characterised accessions underwent a multivariate analysis based on the mean Euclidean distance (dii') with standardised data [15] based on the traits pod length and width, seeds per pod and 100 seed weight with the aid of the program Genes [5].

The 100 seed weight provided the greatest contribution to the divergence between the accessions (99.50%), which demonstrate its importance in studies of diversity among lima bean genotypes. The trait seeds per pod provided the lowest contribution (0.01%) (see Table 2).

Table 2. Relative contribution of the traits pod length, seeds per pod, pod width and 100 seed weight used in the evaluation of genetic divergence in 44 lima bean accessions (Teresina, PI, 2005/2006)

Trait	Value (%)
Pod length (cm)	0.45
Pod width (cm)	0.03
Seeds per pod	0.01
100 bean weight (g)	99.51

The limits of the maximum and minimum distance were found in accessions UFPI-216 (23) and UFPI-155 (44), and UFPI 242 (35) and UFPI 244 (37) (Table 3).

Table 3. Limits of distances from the dissimilarity measures between pairs of lima bean accessions (Teresina, PI, 2005/2006)

Accessions	Maximum distance		Minimum distance	
	Value	Accession pair	Value	Accession pair
01	2.44	01-23	0.33	01-07
02	2.51	02-23	0.43	02-43
03	2.13	03-23	0.33	03-43
04	2.45	04-23	0.29	04-07
05	2.12	05-23	0.20	05-12
06	3.02	06-23	0.46	06-09
07	2.31	07-23	0.29	07-04
08	1.79	08-23	0.31	08-21
09	2.98	09-23	0.15	09-44
10	2.22	10-25	0.17	10-18
11	2.57	11-23	0.43	11-01
12	2.27	12-23	0.20	12-05

Table 3. (Continued)

Accessions	Maximum distance		Minimum distance	
	Value	Accession pair	Value	Accession pair
13	1.76	13-23	0.42	13-14
14	1.71	14-44	0.42	14-13
15	1.69	15-44	0.50	15-16
16	1.71	16-23	0.42	16-14
17	2.37	17-25	0.18	17-18
18	2.35	18-25	0.17	18-10
19	2.63	19-23	0.51	19-02
20	1.95	20-23	0.49	20-16
21	1.89	21-23	0.31	21-08
22	1.94	22-17	0.38	22-32
23	3.10	23-44	0.31	23-35
24	2.67	24-44	0.15	24-33
25	2.37	25-17	0.20	25-32
26	2.36	26-44	0.13	26-39
27	1.87	27-44	0.32	27-29
28	2.30	28-44	0.40	28-30
29	2.00	29-44	0.21	29-36
30	2.17	30-44	0.31	30-40
31	2.35	31-44	0.34	31-36
32	2.29	32-17	0.20	32-25
33	2.67	33-44	0.15	33-24
34	2.60	34-44	0.32	34-24
35	2.89	35-44	0.13	35-37
36	2.17	36-44	0.21	36-29
37	2.97	37-44	0.13	37-35
38	2.07	38-17	0.36	38-41
39	2.35	39-44	0.13	39-26
40	2.16	40-44	0.31	40-30
41	2.08	41-17	0.27	41-36
42	2.30	42-44	0.45	42-28
43	2.14	43-25	0.33	43-03
44	3.10	44-23	0.15	44-09

The UFPI-216 (23) accession from the Vegetable Germplasm Bank of the Federal University of Viçosa (Banco de Germoplasma de Hortaliças da Universidade Federal de Viçosa - BGH-UFV) has black seeds, whereas the accession UFPI-155 (44), which was collected in the northeast of the country, has a cream-coloured seed with a brown hilum. Accessions UFPI-242 (35) and UFPI-244 (37) have brown seeds and are also from BGH-UFV.

Table 4. Entry code of lima bean accessions in the Active Bank at the Federal University of Piauí

Accessions	Code	Accessions	Code	Accessions	Code
1	UFPI – 292	41	UFPI - 346	81	UFPI - 388
2	UFPI - 294	42	UFPI - 348	82	UFPI - 389
3	UFPI - 296	43	UFPI - 349	83	UFPI - 390
4	UFPI - 299	44	UFPI - 350	84	UFPI - 391
5	UFPI - 300	45	UFPI - 351	85	UFPI - 392
6	UFPI - 301	46	UFPI - 353	86	UFPI - 393
7	UFPI - 303	47	UFPI - 354	87	UFPI - 394
8	UFPI - 304	48	UFPI - 355	88	UFPI - 396
9	UFPI - 307	49	UFPI - 356	89	UFPI - 397
10	UFPI - 309	50	UFPI - 357	90	UFPI - 399
11	UFPI - 310	51	UFPI - 358	91	UFPI - 400
12	UFPI - 311	52	UFPI - 359	92	UFPI - 401
13	UFPI - 315	53	UFPI - 360	93	UFPI - 402
14	UFPI - 316	54	UFPI - 361	94	UFPI - 403
15	UFPI - 318	55	UFPI - 362	95	UFPI - 404
16	UFPI - 319	56	UFPI - 363	96	UFPI - 405
17	UFPI - 320	57	UFPI - 364	97	UFPI - 406
18	UFPI - 321	58	UFPI - 365	98	UFPI - 407
19	UFPI - 322	59	UFPI – 366	99	UFPI - 408
20	UFPI - 323	60	UFPI - 367	100	UFPI - 515
21	UFPI - 324	61	UFPI - 368	101	UFPI - 516
22	UFPI - 325	62	UFPI - 369	102	UFPI - 517
23	UFPI - 326	63	UFPI - 370	103	UFPI - 518
24	UFPI - 328	64	UFPI - 371	104	UFPI - 519
25	UFPI - 330	65	UFPI - 372	105	UFPI - 520
26	UFPI - 331	66	UFPI - 373	106	UFPI - 521
27	UFPI - 332	67	UFPI - 374	107	UFPI - 522
28	UFPI - 333	68	UFPI - 375	108	UFPI - 523
29	UFPI - 334	69	UFPI - 376	109	UFPI - 524
30	UFPI - 335	70	UFPI - 377	110	UFPI - 525
31	UFPI - 336	71	UFPI - 378	111	UFPI - 526
32	UFPI - 337	72	UFPI - 379	112	UFPI - 527
33	UFPI - 338	73	UFPI - 380	113	UFPI - 528
34	UFPI - 339	74	UFPI - 381	114	UFPI - 529
35	UFPI - 340	75	UFPI - 382	115	UFPI - 530
36	UFPI - 341	76	UFPI - 383	116	UFPI - 531
37	UFPI - 342	77	UFPI - 384	117	UFPI - 532
38	UFPI - 343	78	UFPI - 385	118	UFPI - 533
39	UFPI - 344	79	UFPI - 386		
40	UFPI - 345	80	UFPI - 387		

In 2006, 118 accessions deposited in the BAGF-UFPI (originally the Soinho Family School of Agriculture [Escola Família Agrícola do Soinho -

EFA-Soinho]) (Table 4) were characterised based on the following morphological descriptors: seed background colour, primary seed colour, secondary seed colour, seed coat pattern, shape and location of the pigments observed on the seed coat, seed shape, seed length and seed width.

After performing the characterisation, a multivariate analysis was conducted to obtain the matrix of dissimilarities with multicategorical variables from the mode of each variable per genotype without replication. Analyses were then performed using the dissimilarity measures by applying the mean Euclidean distance and Tocher grouping, with the help of the computer application Genes [4].

Multivariate analyses are an alternative method of assessing the degree of genetic similarity between treatments, and such analyses summarise a large number of characteristics into a smaller group to facilitate data analysis. Multivariate analyses also indicate the similarities between individuals based on their distances or dispersions plotted in two- or three-dimensional space, and they evaluate the importance of each variable relative to the total variation observed among sampling units so that characteristics that contribute little to the variation can be eliminated [4].

Using a dissimilarity matrix, the hierarchical nearest neighbour method identifies similar progenitors, which are assembled to form the initial group, and the distances of the initial group to other progenitors and groups that formed at later stages are calculated [4]. According to these authors, the Tocher method requires the construction of a dissimilarity matrix, from which the most similar pair of progenitors can be identified. These progenitors form the initial group, and the chance of subsequent inclusion of new progenitors is evaluated. The entry of a progenitor into a group always increases the mean distance value within the group.

Multicategorical data collection is practical, economical and less time consuming compared to quantitative data collection methods. However, each type of data has a unique importance, and it is preferable for a collection to be widely studied so that it can provide increased support to research and value to the collection database [16].

The means of the morphological data (seed length and width) from the accessions were used to obtain a matrix of Euclidean distances, and this matrix was analysed using the hierarchical nearest neighbour method, and a dendrogram was generated (Figure 1).

The hierarchical cluster of accessions was separated vertically at a connection distance of 50% to form four groups.

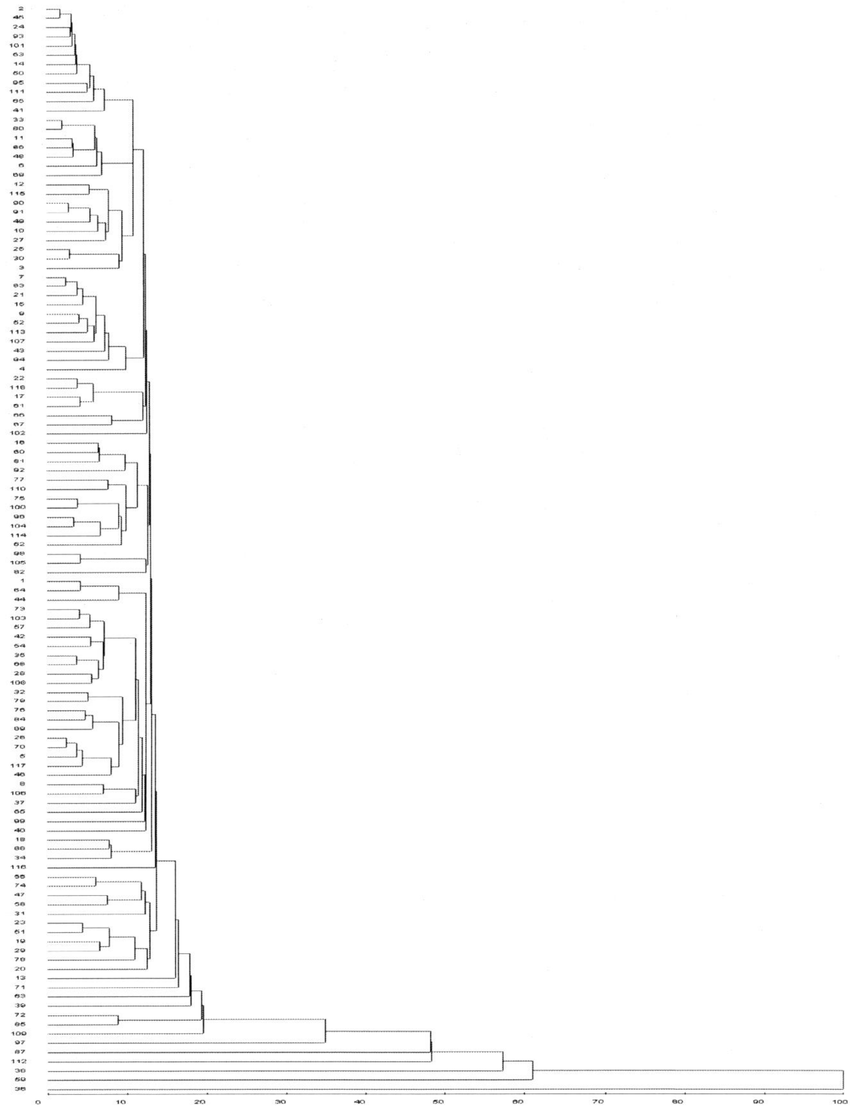

Figure 1. Dendrogram of the genetic dissimilarities among 118 lima bean accessions based on quantitative traits (Teresina, PI, 2006).

Group I included genotype 36, which corresponds to accession UFPI 341 (Table 4) and has the highest values for seed length and width. Group II included genotype 59, which very elongated seeds. Group III included genotype 38, and group IV, which was the largest group, included the remaining 115 genotypes. Group IV was subdivided into two groups, with one including 114 genotypes and the other including only genotype 112. The dendrogram shows the similarity between genotypes 87 and 112, which have similar mean seed length and width values. Through the nearest neighbour analysis and formation of four groups, genetic divergence was observed among the accessions. This divergence is small among the genotypes that constitute Group IV.

The Tocher optimisation analysis showed the formation of five groups (Table 5). Group I, which was the most numerous, consisted of 90 genotypes (76.27%). The second largest group consisted of 21 genotypes (17.79%), and it had the lowest mean seed length and width values. The genotypes allocated to group III (38, 98, 105, 109 and 112) high values for seed length and width, whereas group I had genotypes with medium values, and group II had genotypes with small values.

Group IV was formed solely by genotype 36, and it was completely isolated from the others. The accession contained in the group had the highest quantitative values for seed length and width; this result was consistent with that of group V, which was represented by genotype 59 (UFPI 366) with elongated seeds.

The size of the lima bean seeds can be an important characteristic for the physiological development of the crop because the production of nodules, shoot dry matter and atmospheric nitrogen fixation are associated with size [6].

Table 5. Composition of groupings established by the Tocher method and applied to the standardised mean Euclidean distance matrix among 118 lima bean accessions evaluated with respect to seed length and width (Teresina, PI, 2006)

Group	Individuals (Genotypes)	No. of Genotypes
I	2 45 93 14 101 53 24 50 65 111 41 95 102 43 113 107 9 52 15 116 21 94 7 83 6 86 11 48 80 69 33 10 61 34 90 49 91 17 27 22 118 12 67 115 88 30 25 66 4 18 75 77 110 100 39 3 8 106 114 104 96 62 44 99 16 37 1 81 72 60 85 92 64 97 57 73 103 54 63 40 42 87 82 68 28 35 55 108 71 79	90
II	26 70 5 117 46 32 89 84 76 56 74 58 78 47 13 23 51 20 31 29 19	21
III	98 105 109 38 112	05
IV	36	01
V	59	01

The analyses show that the multivariate techniques (hierarchical nearest neighbour and Tocher optimisation) were consistent. There was similarity in the number of groups formed and allocation of genotypes 36 and 59 to different groups. According to certain authors, the similarity data (genetic divergence) may be useful in breeding programs because it is one of the parameters used in the selection of parents for crosses.

In 2007, a study examined the diversity contained in the BAGF-UFPI using 191 accessions of different origin [2, 3]. The study was conducted at the Laboratory of Plant Genetic Resources of the Department of Biology at the Centre of Natural Sciences, UFPI. The characterisation was based on initial descriptors ("passport" data) and morpho-agronomic descriptors (seed length and width, background colour, primary colour, secondary colour, seed coat pattern, shape and 100 seed weight) published by Biodiversity International [10].

The background colour of the majority of the seeds was white (32.1%), followed by brown (20%) and other colours (8.1%) (Figure 2).

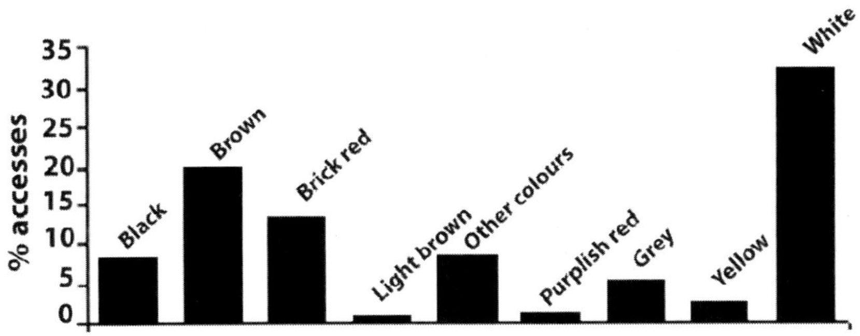

Figure 2. Seed background colour of lima bean accessions at the BAGF-UFPI (Teresina, PI, 2007).

The primary seed colour was largely absent (53%), with smaller representations of light brown (25%), dark brown (14%), and other colours (1.3%) (Figure 3). The secondary seed colour was absent in 86.75% of accessions, and the rest were purplish red (5.30%) and other colours (5.35%) (Figure 4). Most had only one colour.

The pattern of the integument was largely absent, and there was no pigmentation in the seed coat (50.85%). A pigmentation pattern was only observed around the eye (18.6%); distinct eye with many marks on the body and spots over less than half of the body (9.3%); eye with few distinct marks on the body (4%); eye similar to the seed pattern with a spotted hilar region and a few marks, spotted hilar region and front side with a few marks, and moderately mottled body forming blotches (2.6%); and eye similar to the seed pattern with a spotted hilar region and body with bands radiating from the hilar region (1.3%).

The shape and length of the seeds indicate that most of the accessions are from the Hopi branch and include seeds belonging to the Sieva cultigroup, which includes small- to medium-sized seeds with a moderately rounded shape.

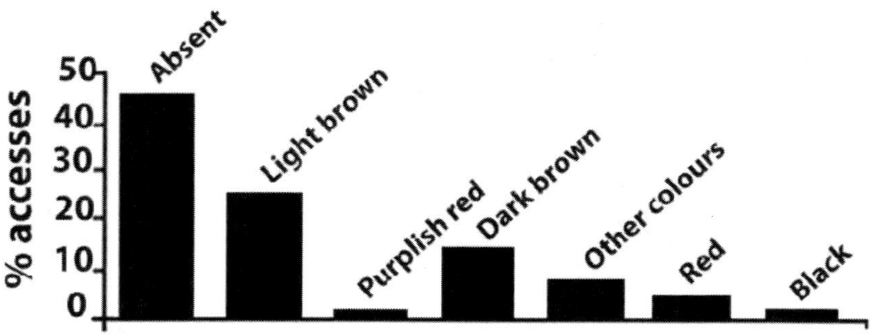

Figure 3. Primary colour of lima bean accessions from the BAGF-UFPI (Teresina, PI, 2007).

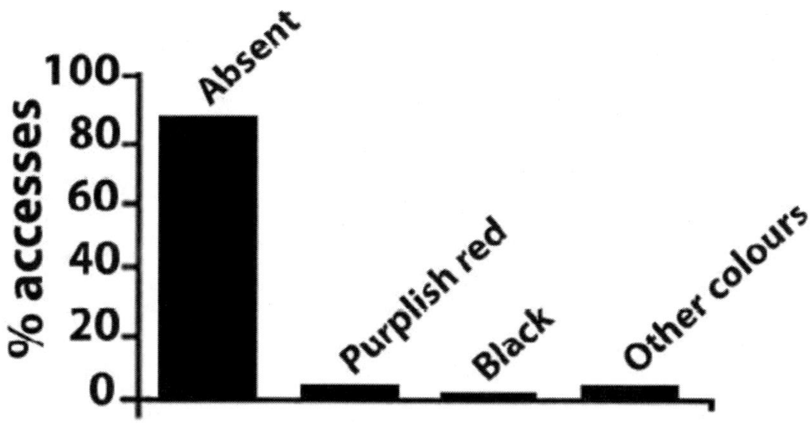

Figure 4. Secondary colour of lima bean accessions from the BAGF-UFPI (Teresina, PI, 2007).

There is variability among the lima bean accessions for all traits; however, the dominant accession included elongated seeds that had a white background colour, no primary and secondary colours, similar eye to the seed pattern, and spotted hilar region (most developed near the micropyle) with marks (Figure 5).

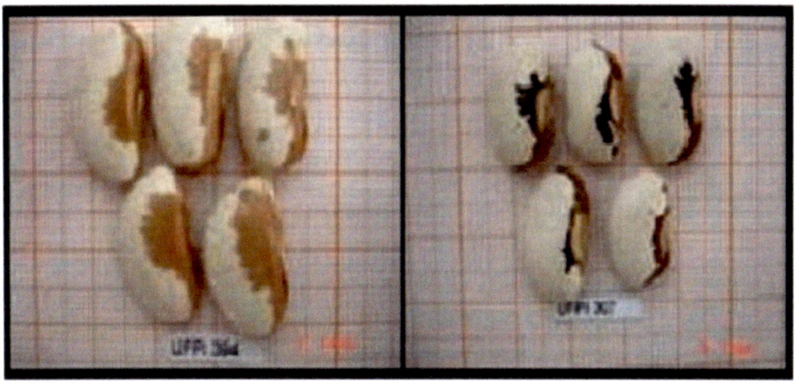

Figure 5. Predominant pattern of seeds at BAGF-UFPI (Teresina, PI, 2007).

After performing the morphological characterisation, dissimilarity analyses of the quantitative traits seed length, seed width and 100 seed weight were conducted by calculating the mean Euclidean distances between accessions. The maximum distance (4.60) was observed between the UFPI-1 and UFPI-503 accessions from BAGF-UFPI and collected in the state of Piauí, and they had 100 seed weights of 73.104 and 25,103 g, respectively. The shortest distance (0.01) was estimated between the accessions UFPI-223 and UFPI-538, which were acquired through exchanges with BGH-UFV in Minas Gerais, and they had 100 seed weights of 35.691 and 35,824 g, respectively. Thus, the maximum distances found in accessions from Piauí indicate that they are the most divergent, and the accessions with smaller distances are less divergent.

The grouping method proposed by Tocher applied to the Euclidean distance matrix was used to divide the 191 accessions into three similarity groups. Group I was the most numerous and contained 181 genotypes representing 94.8% of the total. Group II was composed of accessions UFPI-276, UFPI-479, UFPI-486, UFPI-272, UFPI-289, UFPI-478, UFPI-549, UFPI-46 and UFPI-291; and group III was composed of UFPI-01, which represented 4.2% of the total. This result indicates that UFPI-276, UFPI-479, UFPI-486, UFPI-272, UFPI-289, UFPI-478, UFPI-549, UFPI-469, UFPI-291 and UFPI-01 were genetically divergent from the group I accessions.

The Tocher method did not provide consistent information on the origin of the accessions because accessions of the same origin did not remain in the same group (Table 6). The UFPI-02 and UFPI-26 accessions were allocated to the same group and are from the state of Piauí; however, UFPI-01 is of the same origin and was considered distant from UFPI-02 and UFPI-26.

Most of the accessions from Piauí were sorted into group I; however, Piauí accessions were also present in the other groups, which suggests that there was no relationship between genetic diversity and geographic origin. Similar results were obtained by Abreu [1] with climbing snap beans of indeterminate growth habit from the Germplasm Bank of the North Fluminense State University (Universidade Estadual do Norte Fluminense); these authors found that genotypes of the same origin were not always sorted into the same group. The UENF-1447 and UENF-1448 accessions, which were sorted into the same group, are from the state of Espírito Santo; however, UENF-1440 has the same origin and was considered distant from these two accessions because it had the largest mean values for the number of loci and seeds per pod.

Among the traits that provided the greatest contributions to divergence, the 100 seed weight value explained 97.32% of the genetic divergence. This result was expected because the trait 100 seed weight presented the greatest variation, followed by the length of the seed at 2.02%. In contrast, the width of the seed only contributed 0.66% to the total available variability (Table 7).

The UFPI-01 accession from group III exhibited a 100 seed weight, seed length and seed width of 73.104 g, 17.53 mm and 18.18 mm, respectively. The seed width (18.18 cm) had the most influence on the categorisation of accession UFPI-01 into a separate group, and it presented relatively high distance values in the pairs in which it participated, which produced a satisfactory level of heterozygosis in its progeny. However, in combinations with other accessions, the adaptation of the genotypes must be considered. A similar result was found by Peixoto et al. [13] in an analysis of the genetic divergence among 20 genotypes of snap beans of indeterminate growth habit from the International Centre for Tropical Agriculture (Centro Internacional de Agricultura Tropical - CIAT); Tocher grouping produced four groups, with genotype Hav 49 isolated from the others in a single group.

The clustering produced by the Tocher optimisation technique and mean Euclidean distance suggested that genetic variability occurs among the 191 lima bean accessions in the Active Germplasm Bank of the UFPI, and the 100 seed weight provides the greatest contribution to genetic divergence.

Table 6. Origin of the BAGF-UFPI lima bean accessions and groups formed by Tocher optimisation based on the traits seed length, seed width and 100 seed weight (Teresina, PI, 2007)

Group	Accession	Origin of accession	No. of accessions
I	UFPI-02, UFPI-26, UFPI-32, UFPI-36, UFPI-72, UFPI-459, UFPI-482, UFPI-483, UFPI-491, UFPI-509, UFPI-567, UFPI-459, UFPI-34, UFPI-76, UFPI-463, UFPI-504, UFPI-109, UFPI-112, UFPI-115, UFPI-121, UFPI-286, UFPI-457, UFPI-458, UFPI-554, UFPI-122, UFPI-281, UFPI-293, UFPI-411, UFPI-412, UFPI-414, UFPI-416, UFPI-417, UFPI-129, UFPI-470, UFPI-471, UFPI-472, UFPI-474, UFPI-475, UFPI-476, UFPI-497, UFPI-498, UFPI-499, UFPI-546, UFPI-132, UFPI-495, UFPI-144, UFPI-187, UFPI-288, UFPI-507, UFPI-508, UFPI-487, UFPI-490, UFPI-491, UFPI-154, UFPI-165, UFPI-166, UFPI-167, UFPI-273, UFPI-505, UFPI-506, UFPI-534, UFPI-168, UFPI-189, UFPI-279, UFPI-202, UFPI-205, UFPI-464, UFPI-214, UFPI-265	Piauí	181
	UFPI-274, UFPI-275, UFPI-278, UFPI-456, UFPI-285	Paraíba	
	UFPI-134, UFPI-177, UFPI-178, UFPI-290, UFPI-541, UFPI-568, UFPI-141, UFPI-500, UFPI-501	Ceará	
	UFPI-280, UFPI-282, UFPI-409, UFPI-540, UFPI-555, UFPI-283, UFPI-465, UFPI-467, UFPI-492, UFPI-493, UFPI-494, UFPI-91, UFPI-73, UFPI-468, UFPI-473	Maranhão	
	UFPI-100, UFPI-462	Pernambuco	
	UFPI-216, UFPI-217, UFPI-218, UFPI-220, UFPI-221, UFPI-222, UFPI-223, UFPI-224, UFPI-228, UFPI-229, UFPI-230, UFPI-231, UFPI-232, UFPI-233, UFPI-234, UFPI-235, UFPI-236, UFPI-237, UFPI-238, UFPI-239, UFPI-242, UFPI-243, UFPI-244, UFPI-247, UFPI-250, UFPI-251, UFPI-252, UFPI-253, UFPI-255, UFPI-256, UFPI-257, UFPI-262, UFPI-264, UFPI-419, UFPI-513, UFPI-514, UFPI-538, UFPI-539, UFPI-553, UFPI-556, UFPI-557, UFPI-558, UFPI-559, UFPI-560, UFPI-561, UFPI-562, UFPI-563, UFPI-564, UFPI-565, UFPI-566	BGH – UFV	
II	UFPI-276	Paraíba	09
	UFPI-479	Unknown origin	
	UFPI-486, UFPI-272, UFPI-289, UFPI- 478, UFPI-549	Piauí	
	UFPI-469	Maranhão	
	UFPI-291	Ceará	
III	UFPI-01	Piauí	01

Table 7. Relative contribution of seed length, seed width and 100 seed weight to divergences among lima bean accessions as proposed by Singh (1981) (Teresina, PI, 2007)

Trait analysed	Maximum	Minimum	Mean	Value (%)
100 seed weight (g)	102.27	23.28	56.06	97.32
Seed length (mm)	19.94	8.72	14.23	2.02
Seed width (mm)	18.18	6.60	9.98	0.66

Several authors have recommended the use of parents with maximum divergence to maximise the probability of superior segregants in advanced generations and increase the genetic base. Thus, in crosses in breeding programs, parents must be selected from divergent accessions that also exhibit superior performance in the key characteristics of agronomic importance [4, 7, 12].

In 2008, native lima bean cultivar seeds were collected from farmers, cultivation areas, fairs and markets in municipalities of the state of Piauí, and the coordinates of the corresponding crop field locations were recorded using a Global Positioning System (GPS) receiver.

The collected accessions were registered in the BAGF-UFPI with the following passport data: accession number, common name, collector, collection date and place, colour, and seed number. Subsequently, these accessions were characterised morphologically at the Department of Agronomy/Centre for Agricultural Sciences/UFPI using the list of descriptors published by Biodiversity International [10] for *Phaseolus lunatus* L., and these descriptors are based on the following morphological, qualitative and quantitative seed traits: background colour; primary colour; secondary colour; seed coat pattern; shape, length, width and 100 seed weight.

The divergence between accessions was determined using the Tocher grouping method, with the mean Euclidean distance as the dissimilarity measure. The correlations between the geographical coordinates of origin and genetic distance between lima bean accessions were estimated by Pearson's correlation coefficient. The statistical and genetic analyses were performed using the computer application Genes [5].

The collections of accessions were restricted to regions that effectively produced lima beans in the state of Piauí, i.e., regions that were under constant production between 1990 and 2006 according to data from the Brazilian Institute of Geography and Statistics (Instituto Brasileiro de Geografia e Estatística – IBGE) [9] (Table 8).

Table 8. Native lima bean cultivars, their respective municipalities and localities of occurrence in the state of Piauí and corresponding latitudes and longitudes (Piauí, 2008)

Area	Native cultivar	Municipality/Locality	Latitude	Longitude
1	Boca de Moça	Palmeirais - Cafundó	5° 58'	43° 4'
2	Boca de Moça	Hugo Napoleão	05°59'19"	42°33'22"
3	Boca de Moça	Palmeirais - Serra da Solta	5° 58'	43° 4'
4	Boca de Moça	Barro Duro - Mucambo	5° 51'	42° 29'
5	Boca de Moça	Barro Duro - Tomé Village	5° 51'	42° 29'
6	Boca de Moça	Palmeirais – Olho D' Água	5° 52' 41.8"	42° 34' 40.1"
7	Boca de Moça	Barro Duro – Tomé Village	5° 51'	42° 29'
8	Boca de Moça	Água Branca	05° 53' 24"	42° 38' 16"
9	Boca de Moça	Palmeirais – Olho D' Água	5° 52' 41.8"	42° 34' 40.1"
10	Boca de Moça	Barro Duro – Market	5° 51'	42° 29'
11	Boca de Moça	Palmeirais – Calumbi Farm	5° 58'	43° 4'
12	Boca de Moça	Palmeirais – Serrinha	5° 58' 27.3"	42° 28' 39.1"
13	Boca de Moça	Unknown – Market		
14	Boca de Moça	Bom Principio	6° 31' 44.2"	42° 18' 46.1"
15	Boca de Moça	Palmeirais – Riachinho	5° 58'	43° 4'
16	Boca de Moça	Palmeirais – Riachinho	5° 58'	43° 4'
17	Boca de Moça	Fava Palmeirais – Braço	5° 58'	43° 4'
18	Boca de Moça	Palmeirais – Campestre	5° 58'	43° 4'
19	Boca de Moça	Palmeirais – Fruteira Settlement	5° 58'	43° 4'
20	Boca de Moça	Várzea Grande – Market A	6° 34'	42° 13'
21	Boca de Moça	Várzea Grande – Market B	6° 34'	42° 13'
22	Boca de Moça	Palmeirais – Fruteira Settlement	5° 58'	43° 4'
23	Boca de Moça	Várzea Grande – Central Market	6° 34'	42° 13'
24	Boca de Moça	Barra D' Alcântara – Por Enquanto	6° 32' 17.1"	42° 8' 5.4"
25	Boca de Moça	Lagoa das Pedras – Tanque	6° 38' 51.6"	42° 12' 34.9"
26	Boca de Moça	Regeneração – Piçarra	6° 14'	42° 41'
27	Boca de Moça	Palmeirais – Campestre	5° 58'	43° 4'
28	Boca de Moça	Picos – Market	7° 4'	41° 28'
29	Boca de Moça	Itaueira – Market	7° 36'	43° 2'
30	Boca de Moça	Picos – Market	7° 4'	41° 28'
31	Boca de Moça	Palmeirais – Campestre	5° 58'	43° 4'
32	Boca de Moça	Tanque – Piçarra	06°36'00"	42°16'56"
33	Boca de Moça	Elesbão Veloso - Market	6° 13'	42° 8'
34	Boca de Moça	Picos – Market	7° 4'	41° 28'
35	Boca de Moça	Bom Principio – Carrapato	6° 31' 44.2"	42° 18' 46.1"
36	Boca de Moça	Palmeirais – Dendê	5° 58'	43° 4'
37	Boca de Moça	Novo Oriente – Caraíbas	6° 33' 27"	42° 1' 30.3"
38	Boca de Moça	Barro Duro – Brejão	5° 51'	42° 29'
39	Boca de Moça	Barro Duro	5° 51'	42° 29'
40	Boca de Moça	Elesbão Veloso – Market	6° 13'	42° 8'
41	Boca de Moça	Novo Oriente – Piçarra	6° 33' 27"	42° 1' 30.3"
42	Boca de Moça	Olho D' Água – Lajes - Palmeirais	5° 49.1' 12.4"	42° 34' 11.4"
43	Boca de Moça	Olho D' Água – Lajes - Palmeirais	5° 49.1' 12.4"	42° 34' 11.4"
44	Boca de Moça	Olho D' Água – Lajes -Palmeirais	5° 49.1' 12.4"	42° 34' 11.4"
45	Boca de Moça	Novo Oriente – Caraíbas	6° 33' 27"	42° 1' 30.3"

Table 8. (Continued)

Area	Native cultivar	Municipality/Locality	Latitude	Longitude
46	Boca de Moça	Lageiro – Várzea Grande	6° 36' 33"	42° 9' 26"
47	Boca de Moça	Vaquejador – Várzea Grande	6° 34' 4.6"	42° 12' 46"
48	Boca de Moça	Barra D' Alcântara	6° 31' 23.2"	42° 6' 25.4"
49	Boca de Moça	Santa Rita – Olho D' Água	5° 52' 41.8"	42° 34' 40.1"
50	Boca de Moça	Antero - Palmeirais	5° 58'	43° 4'
51	Boca de Moça	Bom Principio	6° 31' 44.2"	42° 18' 46.1"
52	Boca de Moça	Santa Rita – Olho D'Água	5° 52' 41.8"	42° 34' 40.1"
53	Boca de Moça	Barro Duro	5° 51'	42° 29'
54	Cream	Santa Rita – Olho D' Água	5° 52' 41.8"	42° 34' 40.1"
55	White	Bom Jesus	9° 4'	44° 21'
56	White	Ponta D'Água – Palmeirais	5° 58'	43° 4'
57	White	José de Freitas –Ema Community	4° 45'	42° 35'
58	White	José de Freitas – Market	4° 45'	42° 35'
59	Cream	Palmeirais - Estamos Unidos	5° 58'	43° 4'
60	White	Picos – Market	7° 4'	41° 28'
61	White	Landri Sales – Market	7° 16'	43° 56'
62	White	Itaueira – Market	7° 36'	43° 2'
63	White	São Miguel do Tapuio – Market	5° 30'	41° 20'
64	Cream	Picos – Market	7° 4'	41° 28'
65	Cream	José de Freitas – Market	4° 45'	42° 35'
66	Pink	José de Freitas - Market	4° 45'	42° 35'
67	Cream	Itaueira – Market	7° 36'	43° 2'
68	Grey	José de Freitas – Carimã Settlement	4° 45'	42° 35'
69	Grey	José de Freitas – Carimã Settlement	4° 45'	42° 35'
70	White	José de Freitas – Carimã Settlement	4° 45'	42° 35'
71	Cream	Hugo Napoleão	05°59'19"	42°33'22"

For the correlation analysis, the areas in similar locations were grouped (Figure 6). A total of 24 groups were formed, with group 1 including the largest number of areas (15 - Table 9). Groups 5, 6, 9, 10, 11, 14, 18, 19, 20, 21, 23 and 24 only included one area each. Area 13 was not included in any group because it was not possible to determine its geographic coordinates. The genetic distances between the groups ranged from 2.79 to 0.13 (Table 10).

The smallest genetic distance was found between groups 19 and 20, and the highest distance was found between groups 9 and 24. The geographic distances between collection groups ranged from 516.3 to 0.5 km (Table 11).

The smallest geographic distance was found between groups 8 and 19, and the largest distance was found between groups 21 and 24. The matrix of genetic distances between groups was used to construct the dendrogram shown in Figure 7.

Table 9. Representation of groups, areas contained in the group and geographic coordinates (longitude and latitude) of each group in the state of Piauí

Group	Areas contained in the group	Latitude	Longitude
1	1,3,11,15,16,17,18,19,22,27,31,36,50,56,59	5° 58'	43° 4'
2	2,71	05°59'19"	42°33'22"
3	4,5,7,10,38,39,53	5° 51'	42° 29'
4	6,9,49,52,54	5° 52' 41.8"	42° 34' 40.1"
5	8	05° 53' 24"	42° 38' 16"
6	12	5° 58' 27.3"	42° 28' 39.1"
7	14,35,51	6° 31' 44.2"	42° 18' 46.1"
8	20,21,23	6° 34'	42° 13'
9	24	6° 32' 17.1"	42° 8' 5.4"
10	25	6° 38' 51.6"	42° 12' 34.9"
11	26	6° 14'	42° 41'
12	28,30,34,60,64	7° 4'	41° 28'
13	29,62,67	7° 36'	43° 2'
14	32	06°36'00"	42°16'56"
15	33,40	6° 13'	42° 8'
16	37,41,45	6° 33' 27"	42° 1' 30.3"
17	42,43,44	5° 49.1' 12.4"	42° 34' 11.4"
18	46	6° 36' 33"	42° 9' 26"
19	47	6° 34' 4.6"	42° 12' 46"
20	48	6° 31' 23.2"	42° 6' 25.4"
21	55	9° 4'	44° 21'
22	57,58,65,66,68,69,70	4° 45'	42° 35'
23	61	7° 16'	43° 56'
24	63	5° 30'	41° 20'

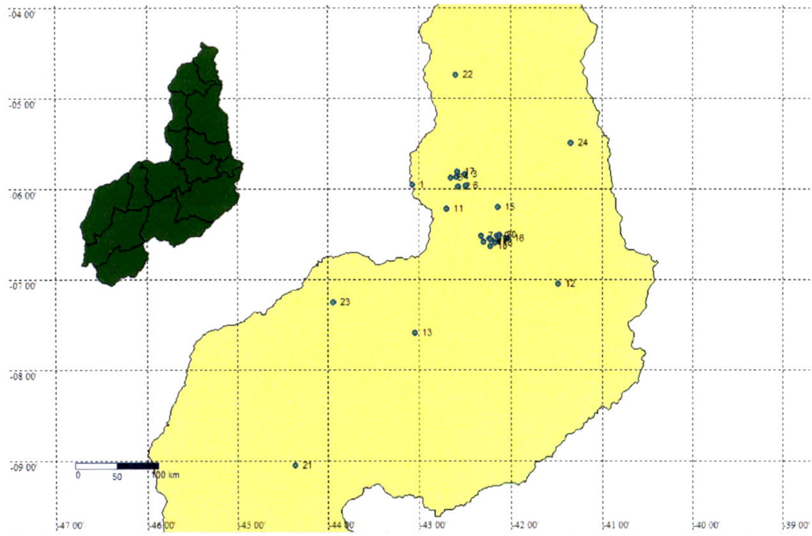

Figure 6. Representation of the groups in the study of lima bean crops visited in the state of Piauí.

Table 10. Matrix of genetic distances between the 24 groups of native lima bean cultivars in the state of Piauí

	1	2	3	4	5	6	7	8	9	10	11	12	13	14	15	16	17	18	19	20	21	22	23
1	-																						
2	1.68	-																					
3	0.71	1.51	-																				
4	1.28	1.62	1.16	-																			
5	0.55	1.72	0.62	1.24	-																		
6	0.62	1.52	0.51	1.09	0.53	-																	
7	0.64	1.57	0.61	1.16	0.43	0.50	-																
8	0.70	1.52	0.62	1.21	0.51	0.58	0.59	-															
9	0.71	2.05	0.88	1.54	0.37	0.89	0.74	0.81	-														
10	0.73	2.07	0.86	1.54	0.38	0.83	0.75	0.83	0.20	-													
11	1.02	1.30	0.80	1.10	1.18	0.77	0.85	0.97	1.55	1.54	-												
12	1.00	1.65	0.96	1.42	0.82	0.94	0.90	0.89	1.03	1.05	1.22	-											
13	1.71	1.46	1.56	1.63	1.79	1.55	1.59	1.57	2.13	2.15	1.22	1.64	-										
14	0.76	1.37	0.56	0.98	0.85	0.38	0.61	0.74	1.22	1.18	0.39	1.06	1.37	-									
15	0.58	1.65	0.58	1.21	0.19	0.45	0.46	0.46	0.49	0.45	1.10	0.83	1.72	0.76	-								
16	0.59	1.46	0.47	1.07	0.47	0.19	0.47	0.48	0.83	0.78	0.77	0.88	1.51	0.42	0.37	-							
17	0.68	1.53	0.56	1.11	0.56	0.35	0.55	0.62	0.89	0.82	0.76	0.97	1.57	0.44	0.47	0.33	-						
18	0.59	1.69	0.61	1.22	0.27	0.34	0.42	0.55	0.60	0.56	1.06	0.87	1.72	0.69	0.30	0.36	0.46	-					
19	0.72	1.34	0.53	1.02	0.80	0.42	0.54	0.67	1.17	1.16	0.40	1.01	1.34	0.19	0.74	0.43	0.49	0.67	-				
20	0.66	1.33	0.47	1.02	0.68	0.37	0.50	0.56	1.05	1.05	0.51	0.93	1.36	0.26	0.62	0.33	0.46	0.58	0.13	-			
21	1.36	1.32	1.18	1.36	1.56	1.26	1.26	1.22	1.89	1.91	0.58	1.38	1.07	0.92	1.48	1.21	1.22	1.51	0.87	0.93	-		
22	1.68	1.69	1.55	1.75	1.77	1.61	1.59	1.57	2.00	2.03	1.30	1.70	1.61	1.42	1.73	1.57	1.60	1.75	1.34	1.36	1.24	-	
23	1.62	1.37	1.44	1.54	1.82	1.51	1.50	1.48	2.15	2.19	0.78	1.56	1.04	1.16	1.76	1.48	1.49	1.75	1.09	1.17	0.38	1.32	-
24	2.23	1.33	2.03	1.81	2.44	2.05	2.13	2.11	2.79	2.78	1.31	1.98	1.13	1.68	2.34	2.04	2.02	2.35	1.71	1.79	0.96	1.55	0.88

Table 11. Matrix of geographic distances between the 24 groups of native lima bean cultivars in the state of Piauí

1	2	3	4	5	6	7	8	9	10	11	12	13	14	15	16	17	18	19	20	21	22	23
-																						
56.6	-																					
65.9	17.3	-																				
55.0	12.4	10.9	-																			
48.3	14.2	17.7	6.8	-																		
65.3	8.8	13.8	15.4	20.0	-																	
104.1	65.5	77.4	77.7	79.3	64.0	-																
115.2	74.1	84.6	86.0	88.2	71.6	11.4	-															
120.9	76.6	85.3	87.9	90.7	73.0	19.7	9.6	-														
121.1	82.3	93.3	94.3	96.3	80.2	17.4	9.0	14.7	-													
51.7	30.5	47.8	41.0	38.3	36.6	52.4	63.4	69.4	69.6	-												
214.7	169.5	175.4	179.9	183.6	164.6	110.8	99.7	94.2	94.3	163.1	-											
180.7	185.8	202.9	197.0	194.1	190.0	142.7	145.6	153.8	139.2	156.0	182.8	-										
111.5	74.1	85.9	86.3	87.8	72.5	8.6	8.1	17.7	9.6	60.1	103.9	138.3	-									
107.0	53.2	56.1	61.8	66.5	46.6	39.8	39.8	35.5	48.4	60.9	119.5	182.5	45.5	-								
132.5	86.1	93.2	96.9	100.2	81.7	32.0	21.2	12.3	22.7	81.2	83.6	160.3	28.8	39.5	-							
57.4	18.7	10.1	6.5	10.8	19.9	83.4	91.3	92.9	99.8	47.4	184.1	203.4	91.9	65.3	101.4	-						
123.2	81.6	91.4	93.3	95.7	78.7	19.4	8.1	8.2	7.2	71.5	91.6	146.2	13.9	43.5	15.7	98.5	-					
115.5	74.5	84.9	86.3	88.5	71.9	11.9	0.5	9.2	8.8	63.9	99.2	145.8	8.5	39.8	20.8	91.7	7.7	-				
122.7	77.2	85.3	88.3	91.4	73.2	22.8	13.0	3.5	17.8	71.4	92.9	157.0	21.2	34.0	9.8	93.1	11.0	12.7	-			
370.9	393.8	411.1	403.2	398.9	399.6	359.5	363.0	371.3	356.8	363.3	387.2	217.6	355.6	398.9	377.8	409.3	363.8	363.2	374.6	-		
144.8	137.0	122.1	124.8	126.2	135.9	199.0	205.0	203.9	213.9	164.4	284.5	319.1	207.3	169.7	209.2	118.3	210.9	205.2	203.1	515.8	-	
172.8	207.8	224.2	214.6	209.1	215.3	196.8	204.9	214.5	202.4	179.3	273.4	106.0	196.8	230.4	225.0	219.8	209.3	205.2	217.9	204.3	315.9	-
198.8	145.8	133.1	144.1	150.8	137.2	157.2	153.2	145.1	159.7	170.1	173.9	298.8	160.7	118.9	139.8	141.5	152.9	153.1	141.9	516.3	161.5	347.7

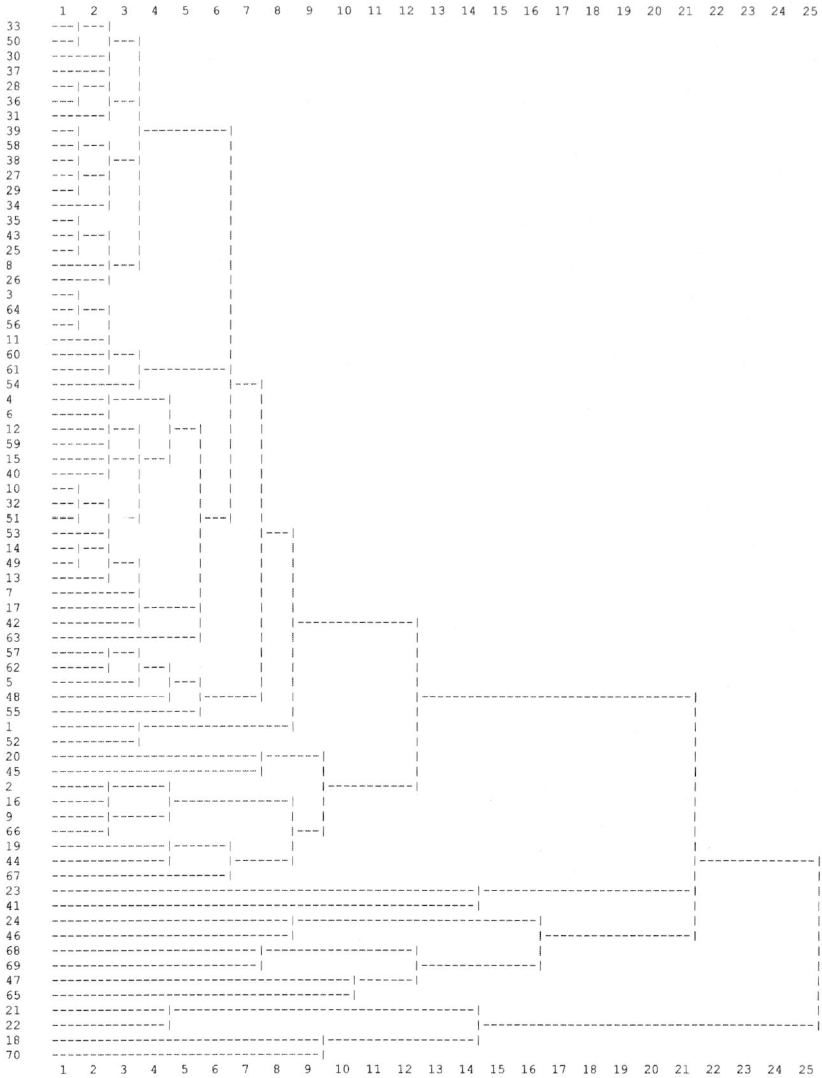

Figure 7. Dendrogram of genetic distances between the 24 native cultivars of lima bean in the state of Piaui.

There was a positive correlation of 0.4109, and significance was determined at a p-value of 1% based on the t-test between genetic distance and geographic distance (Table 12).

Table 12. Estimates of the variance of genetic and geographic distances and the covariance and correlations between the geographic and genetic distance of the groups studied

No. of data	Variance (geographic distance)	Variance (genetic distance)	Covariance	Correlation
552	10830.3415	0.2977	23.3309	0.4109**

**: significant at a p-value of 1% according to the t test.

FINAL CONSIDERATIONS

Morphological characterisations can provide valuable information on the genetic variability of accessions, and such data can assist in the characterisation of germplasm and provide descriptions of the genetic divergence between genotypes.

The characterisation of seeds using quantitative and qualitative descriptors is important in studies of genetic diversity among accessions of lima beans collected in the central northern region of Brazil because it clusters the accessions into similar groups. Such characterisations are economical, easy to use and efficient at establishing variability among accessions. These characterisations also have the potential for use in the selection of desirable genotypes for breeding programs destined for the consumer market, which has variable preferences for grains based on the size and colour, which is specific to each region.

Genetic variability was observed among the native lima bean cultivars in the state of Piauí, and a trend towards regionalisation was also observed, indicating that there is little exchange of materials between groups in the areas studied.

REFERENCES

[1] Abreu, F.B., Leal N.R., Rodrigues R., Amaral J.A.T. and Silva D.J.H., Genetic divergence among climbing snap bean accessions. *Horticultura Brasileira* [Serial on the Internet]. 2004 Sep [cited 2006 Aug 26]; 22(3): 547-552. Available from: http: //www.scielo.br/scielo.php

[2] Costa, E.M.R., Lopes, A.C.A., Gomes, R.L.F., Silva, R.N.O., Silva, S.C.C.C. Characterisation of lima bean accessions in the Germplasm

Bank of the Federal University of Piauí [Caracterização de Acessos de Fava do Banco de Germoplasma da Universidade Federal do Piauí]. *4º Congresso Brasileiro de Melhoramento de Plantas,* São Lourenço, MG. P.1-3, 2007a.

[3] Costa, E.M.R., Santos, J.O., Silva, S.C.C.C., Lopes, A.C.A., Gomes, R.L.F., Dissimilarity of lima bean accessions in the Germplasm Bank of the Federal University of Piauí [Dissimilaridade de acessos de fava do Banco de Germoplasma da Universidade Federal do Piauí]. *6º Simpósio Internacional de Recursos Genéticos para a América Latina e Caribe,* Chapingo, México, 2007b, p.43.

[4] Cruz, C.D., Regazzi, A.J., Carneiro, P.C.S. *Biometric models applied to genetic improvement* [*Modelos biométricos aplicados ao melhoramento genético*]. v.1, 3. ed. Viçosa, MG: UFV, 480 p. 2004.

[5] Cruz, C.D., *Genes Program: Windows version - Computational Application in Statistical Genetics* (*Programa Genes: versão Windows – Aplicativo computacional em genética e estatística).* Viçosa: Editora UFV, 648p. 2006.

[6] Dobert, R.C. and Blevins, D.G., Effect of seed size and plant growth on nodulation and nodule development in lima bean *(Phaseolus lunatus* L.). *Plant and Soil* 148: 11-19, 1993.

[7] Ferreira, M.E. and Grattapaglia, D., *Introduction to the use of molecular markers in genetic analysis* [*Introdução ao uso de marcadores moleculares em análises genéticas*]. 3. ed. Brasília: EMBRAPA-CENARGEM, 1998. 220 p.

[8] Hamrick, J.L. and Godt, J.W., Allozyme diversity in plant species. *In* A. H. D. Brown, M. T. Clegg, A. L. Kalher, and B. S. Weir [eds.], *Plant population genetics, breeding, and genetic resources,* 43–63. Sinauer Associates, Sunderland, Massachusetts, USA, 1990.

[9] IBGE. 2006. Sistema IBGE de Recuperação Automática (SIDRA): Municipality Agricultural Database [Banco de Dados: Produção Agrícola Municipal], Available from: Http://www.sidra.ibge.gov.br/bda/tabela/protabl1.asp?z=t&o=10&i=P. Accessed on 29/01/2007.

[10] IPGRI. Descriptors for *Phaseolus lunatus* (lima bean) [Descritores para *Phaseolus lunatus* (feijão-espadinho)]. International Plant Genetic Resources Institute, Rome. 2001. 51p.

[11] Lemos, L.B., Oliveira, R.S., Palomino, E.C., Silva, T.R.B, Agronomic and technological characteristics of genotypes of the Carioca commercial bean group [Características agronômicas e tecnológicas de

genótipos de feijão do grupo comercial Carioca]. *Pesquisa Agropecuária Brasileira*, v.39, n.4, p.319-326, 2004.

[12] Maluf, W.R. and Ferreira, P.E., Multivariate analysis of genetic divergence in snap beans (*Phaseolus vulgaris* L.) [Análise multivariada da divergência genética em feijão-de-vagem (*Phaseolus vulgaris* L.)]. *Horticultura Brasileira*, Brasília, v.1, n.2, p.31-34. 1983.

[13] Peixoto, N., Braz, L.T., Banzatto, D.A., Moraes, E.A., Moreira, F.M., Agronomic characteristics, productivity, quality of pods and genetic divergence in indeterminate growth snap beans [Características agronômicas, produtividade, qualidade de vagens e divergência genética em feijão-vagem de crescimento indeterminado]. *Horticultura Brasileira,* Brasília, v.20, n.3, p.447-451, 2002.

[14] Santos, D., Corlett, F.M.F., Mendes, J.E.M.F., Wanderley Júnior, J.SA. Productivity and morphology of pods and seeds of beans varieties in the state of Paraíba [Produtividade e morfologia de vagens e sementes de variedades de fava no estado da Paraíba]. *Pesquisa Agropecuária Brasileira,* Brasília, v.37, n.10, p.1407-1412, 2002.

[15] Santos, J.O., Sousa, I.F.S., Gomes, R.L.F., Lopes, A.C.A. Genetic divergence among UFPI lima bean gene bank accessions [Divergência genética entre acessos de fava do Banco de Germoplasma da UFPI]. *4º Congresso Brasileiro de Melhoramento de Plantas,* São Lourenço, MG. P.1-3, 2007c.

[16] Sudré, C.P., Cruz, C.D., Rodrigues, R., Riva, E.M., Amaral Júnior A.T., Silva D.J.H., Pereira T.N.S. Multicategoric variables in determining the genetic divergence among accessions of pepper and chili [Variáveis multicategóricas na determinação da divergência genética entre acessos de pimenta e pimentão]. *Horticultura Brasileira,* 24: 88-93, 2006.

[17] Torggler, M.G.F., Contel, E.P.B., Torggler, S.P., *Isozymes: genetic variability in plants* [*Isoenzimas: variabilidade genética em plantas*]. Ribeirão Preto: SBG, 1995, 186 p.

[18] Valls, F.J.M. Morphological, reproductive and biochemical characterisation of plant germplasm [Caracterização morfológica, reprodutiva e bioquímica de germoplasma vegetal]. In: Meeting On Genetic Resources [Encontro Sobre Recursos Genéticos], *Proceedings.* Jaboticabal: UNESP, 1988. p. 106-128.

[19] Vieira, R.F. Cultivation of lima bean [A cultura do feijão-fava]. *Informe Agropecuário.* Belo Horizonte: EPAMIG, v.16, n.174, p.30-37, out. 1992.

[20] Yeeh, Y., Kang, S.S., Chung, M.G. Evaluation of the natural monument populations of *Camellia japonica* (Thearaceae) in Korea based on allozyme studies. *Botanical Bulletin of Academia Sinica,* Shanghai, v. 37, n. 1, p. 141-146, 1996.

[21] Zoro Bi, I. 1999. Variabilité génétique des populations sauvages de *Phaseolus lunatus* L. dans la vallé centrale du Costa Rica et ses implications dans la mise au point d'une stratégie de conservation *in situ. Ph.D. thesis,* Faculte Universitaire des Sciences Agronomiques, Gembloux, Belgium.

[22] Zoro Bi, I; Maquet, A.; Baudoin, J.P. Population genetic structure of wild *Phaseolus lunatus* (Fabaceae), with special reference to population sizes. *American Journal of Botany* 90(6): 897–904. 2003.

In: Phaseolus Lunatus
Editors: A. S. Ferreira de Araujo et al.
ISBN: 978-1-63482-494-1

Chapter 2

GENETIC VARIABILITY OF SEED CHARACTERS OF LIMA BEAN

Jardel Oliveira Santos[1*], Regina Lucia Ferreira Gomes[2] and Ângela Celis de Almeida Lopes[2]

[1]Universidade Federal do Maranhão,
Boa Vista, Chapadinha, Maranhão, Brasil

[2]Universidade Federal do Piauí, Campus Ministro Petrônio Portela,
s/n, Bairro Ininga, Teresina, Piauí, Brasil

ABSTRACT

The goals of the present study were to perform physical and chemical characterisation of seeds and to estimate the genetic divergence between accessions of lima bean (*Phaseolus lunatus* L). The examined genetic material consisted of 27 accessions obtained from the Active Germplasm Bank of Piauí Federal University (Banco Ativo de Germoplasma da Universidade Federal do Piauí - BAG UFPI) cultivated during the 2006 crop year. Seeds from the selected accessions were evaluated regarding the following chemical characters: the contents of moisture, ash, the ether extract, crude protein, the non-nitrogen extract, total, insoluble and soluble dietary fibre, and hydrogen cyanide. The evaluated physical seed characters were length, width, thickness and 100-seed weight. Based on these characters, the seeds were classified in terms of their shape

* Corresponding Author address: Centro de Ciências Agrárias e Ambientais, BR-222, KM 04, S/N, Boa Vista. CEP 65500-000, Chapadinha-MA, Brazil, E-mail: jardel.os@ufma.br

(spherical, elliptical and oblong/kidney-shaped), profile (flat and semi-flat) and size (small, medium, normal and large). The genetic divergence between accessions was estimated through Tocher's clustering method using the Mahalanobis distance as a measure of dissimilarity, and ten groups were formed. The tested lima bean accessions from BAG UFPI exhibited variability in physical and chemical seed characters. Insoluble and soluble fibre were the characters that contributed most to the total divergence between accessions and enabled the selection of divergent progenitors to obtain segregating populations.

Keywords: Germplasm, protein, hydrogen cyanide and dissimilarity

INTRODUCTION

Lima bean (*Phaseolus lunatus* L.), also known as butter bean and, in Brazil, as 'feijão-fava', fava-belém, 'feijão-espadinho', 'feijão-farinha', 'feijão-fígado-de-galinha', 'feijão-de-lima' or 'fava-de-lima', and appears to show a wider adaptive capacity, as lima bean is also considered more resistant to drought, high temperatures and moisture, developing better in the humid and hot tropics than the common bean (*P. vulgaris* L.) (Rachie et al., 1980). Lima bean cultivation is relatively limited, which is believed to be mainly due to the traditional consumption of common bean and the lack of varieties adapted to the conditions of the regions interested in producing it (Lymman, 1983).

Genetic diversity between lima bean genotypes has been observed for characteristics related to the chemical and physical composition of seeds, and such diversity may be useful in breeding programs (Azevedo et al., 2003; Santos et al., 2011). In addition, characteristics related to the physical, chemical and culinary quality of the grain are important for bean consumers and are considered additional selection criteria in breeding programs (Pérez-Herrera et al., 2002).

Estimates of genetic divergence between genotypes allow inference of the specific combination capacity and heterosis, even before field crossings are performed. This increases the chance of identifying and recovering more promising combinations from segregating populations (Oliveira et al., 2003; Abreu et al., 2004).

The structuring of divergence groups is fundamental for the choice of progenitors. Crossings between genotypes belonging to more distant groups are considered more promising because they are more dissimilar (Santos et al.,

2011). The use of Mahalanobis' generalised distance (D2) as a measure of dissimilarity and of the Tocher and UPGMA clustering methods has been shown to be effective for the identification of genetic diversity (Abreu et al., 2004; Coellho et al., 2007).

The goal of the present study was to estimate the genetic divergence between lima bean accessions based on certain physical and chemical characteristics of the seeds.

METHODS

Twenty-seven lima bean (*Phaseolus lunatus* L.) accessions were used, which were obtained from the Active Germplasm Bank of Piauí Federal University (Banco Ativo de Germoplasma da Universidade Federal do Piauí-BAG UFPI) (Table 1) cultivated during the 2006 crop year.

Physical characterisation of the seeds was performed based on length (C), width (L) and thickness (E) measurements (in millimetres) from ten seeds chosen randomly and the 100-seed weight (weight, in grams, of 100 seeds chosen at random; 100SW). Based on this characterisation, the seeds were classified in terms of their size (according to Mateo Box) and shape and profile (using the J index calculated as the length/width ratio (L/W) and the H index calculated as the thickness/width ratio (T/W), respectively, according to Puerta Romero) (cited in Vilhordo et al., 1996). Chemical characterisation of the seeds was performed using a 200-g seed sample from each lima bean accession.

The samples were ground using an electrical food processor and stored in polyethylene bags in a cool and dry place, at between 19 and 25°C. The processed samples were analysed according to AOAC (2005) procedures, and the following characters were determined: the contents of moisture (MO), ash (ASH), the ether extract (EE), crude protein (CP), the non-nitrogen extract (NNE), neutral detergent fibre (NDF), acid detergent fibre (ADF), hemicellulose (HEM), and hydrogen cyanide (HCN). NNE was calculated according to the following equation: 100 - (moisture + ash + ether extract + crude protein). NDF, representing total dietary fibre (TDF), and ADF, representing insoluble dietary fibre (IDF), were determined according to Van Soest and Wine (cited in Silva, 1990). HEM represented soluble dietary fibre (SDF) and was calculated as NDF - ADF. HCN was quantified using the argentometric method.

Table 1. Lima bean accessions from the Active Germplasm Bank (BAG) of UFPI, according to collection sites and seed coat colour

BAG code	Collection site	Seed coat colour
UFPI-032	Várzea Grande – Piauí	brown
UFPI-121	Bom Jesus – Piauí	off-white
UFPI-134	Ceasa – Ceará	white
UFPI-173	Unknown	striped white and off-white
UFPI-177	Ceasa – Ceará	off-white
UFPI-189	Inhuma – Piauí	white
UFPI-216	Itapiraquan – Goiás	purple
UFPI-220	Ipamari – Goiás	brown
UFPI-222	Ipameri – Goiás	purple
UFPI-228	Porto Firme – Minas Gerais	purple
UFPI-229	Rio Casca – Minas Gerais	purple
UFPI-230	Porto Firme – Minas Gerais	purple
UFPI-234	Coimbra – Minas Gerais	purple
UFPI-237	Bom Sucesso – Minas Gerais	spotted brown
UFPI-243	Rio Casca – Minas Gerais	brown
UFPI-251	Cácere – Mato Grosso	striped off-white and black
UFPI-264	Ipanema – Minas Gerais	black
UFPI-271	Paraguai	white
UFPI-272	Unknown	cream
UFPI-274	Cajazeiras – Paraíba	white
UFPI-275	Cajazeiras – Paraíba	striped black
UFPI-276	Cajazeiras – Paraíba	striped purple and white
UFPI-278	Cajazeiras – Paraíba	off-white
UFPI-280	Maranhão	white
UFPI-282	Fazenda Santa Fé – Maranhão	purple
UFPI-468	São Domingos – Maranhão	cream
UFPI-491	Várzea Grande – Piauí	cream with different hilum diferente

In the multivariate analysis, the genetic divergence between accessions was estimated by the Tocher (Rao, 1952) and UPGMA (Unweighted Pair Group Method with Arithmetic Mean) clustering methods, using the

Mahalanobis distance as a dissimilarity measure. A univariate analysis of variance, meeting all of the required assumptions, was initially applied to the data, and the analysis was conducted under a completely randomised design. The relative contribution of the tested characters to divergence was determined according to Singh (Singh, 1981), supplemented by discarding variables according to Oliveira et al. (2004).

Statistical-genetic analyses were performed using Genes software (Cruz, 2007). Significant differences ($p<0.05$) between the tested lima bean genotypes were observed for all of the evaluated characters (Tables 2 and 3), indicating genetic variability. The minimum and maximum coefficients of variation were 2.17% (non-nitrogen extract) and 21.49% (hemicellulose), respectively.

The lowest dissimilarity between pairs of accessions, determined based on Mahalanobis's generalised distance (D^2), was observed between accessions UFPI-222 and UFPI-278 (D^2 = 21.50), whereas the highest dissimilarity was observed between accessions UFPI-282 and UFPI-491 (D^2 = 294.53). The use of dissimilarity-based multivariate analyses increases the likelihood of recovering superior genotypes from segregating generations (Carvalho et al., 2003), especially by crossing individuals from more distant groups, as assessed by these analyses (Karasawa et al., 2005). In addition to using divergent genotypes, superior performance in regard to the main characters of economic importance should also be considered when recommending crossings (Cruz et al., 2004). Behaviours *per se* should therefore also be taken into account when choosing genotypes (Abreu et al., 2004).

The clustering analysis using the Tocher method resulted in the formation of ten groups. Group 1 included six accessions, two with a high ash content (UFPI-222 and UFPI-229) and four of the accessions showing the highest crude protein concentrations (UFPI-222, UFPI-229, UFPI-243 and UFPI-278), all of which exhibited spherical and semi-flat seeds, except for UFPI-278 (flat profile), and three of the accessions presented a small size (UFPI-222, UFPI-230 and UFPI-237). However, this group also contained five of the accessions displaying the second highest hydrogen cyanide concentration, which is an undesirable character.

Group 2 included seven accessions, showing prevalence of accessions with desirable characters, such as a high ash content (UFPI-234 and UFPI-264), high crude protein content (UFPI-216, UFPI-234, UFPI-251, UFPI-264, UFPI-275 and UFPI-274), lower hydrogen cyanide concentration (UFPI-228, UFPI-251, UFPI-264, UFPI-274 and UFPI-275) and elliptical shape (UFPI-251 and UFPI-275).

Table 2. Mean[1] contents (g/100 g) of moisture (MO), ash (ASH), the ether extract (EE), crude protein (CP), the non-nitrogen extract (NNE), neutral detergent fibre (NDF), acid detergent fibre (ADF), hemicellulose (HEM) and hydrogen cyanide (HCN, mg/kg) in the seeds of the 27 tested lima bean accessions. Teresina, Piauí, 2008

Accession	MO	ASH	EE	CP	NNE	NDF	ADF	HEM	HCN
UFPI-032	8.94e	3.22d	0.67c	17.84b	69.54a	56.43a	10.42b	46.01a	104.42b
UFPI-121	10.41b	3.05e	0.56c	21.34a	64.89c	54.43a	7.08e	47.35a	87.68b
UFPI-134	10.54b	3.39b	0.55c	22.28a	63.38c	33.42c	5.65f	27.77c	47.22c
UFPI-173	9.48d	3.16d	0.60c	16.85b	70.07a	29.83c	6.50e	23.33d	50.54c
UFPI-177	10.39b	2.62g	0.93a	18.55b	68.05b	35.86c	7.37e	28.50c	87.52b
UFPI-189	9.76c	3.03e	0.47d	20.25a	66.33b	45.32b	9.50c	35.82b	92.33b
UFPI-216	8.57f	3.10e	0.54c	20.90a	66.89b	30.27c	8.36d	21.92d	97.08b
UFPI-220	9.81c	3.09e	0.33d	20.67a	65.96b	39.35b	11.88a	27.47c	53.26c
UFPI-222	10.06c	3.57a	0.24e	20.66a	65.19c	45.85b	8.81d	37.04b	82.83b
UFPI-228	9.37d	3.12e	0.63c	19.52b	67.32b	33.13c	10.31b	22.83d	75.38c
UFPI-229	10.79b	3.63a	0.41d	22.98a	62.24c	38.18c	8.59d	29.59c	83.76b
UFPI-230	10.37b	3.16d	0.76b	18.63b	67.18b	40.89b	8.42d	32.46b	117.38b
UFPI-234	8.52f	3.54a	0.59c	21.32a	66.12b	40.57b	10.53b	30.04c	96.64b
UFPI-237	9.80c	3.44b	0.71c	19.68b	66.58b	24.26c	8.44d	15.82d	62.59c
UFPI-243	11.25a	3.15d	0.37d	21.57a	63.58c	41.75b	8.33d	33.43b	84.30b
UFPI-251	9.03e	3.46b	0.34d	20.56a	66.32b	42.10b	8.59d	33.51b	64.60c
UFPI-264	9.42d	3.70a	0.21e	22.27a	64.34c	43.57b	8.36d	35.21b	44.49c
UFPI-271	10.61b	3.40b	0.38d	21.63a	64.00c	52.62a	6.85e	45.77a	160.63a

Accession	MO	ASH	EE	CP	NNE	NDF	ADF	HEM	HCN
UFPI-272	9.79c	2.68g	0.47d	19.20b	67.83b	46.52b	9.51c	37.02b	88.96b
UFPI-274	10.16c	3.25d	0.45d	21.26a	64.77c	40.54b	8.54d	32.00b	58.00c
UFPI-275	8.26f	3.48b	0.62c	21.84a	65.68c	29.98c	8.45d	21.53d	66.02c
UFPI-276	9.09e	3.29c	0.32d	22.65a	64.52c	46.88b	6.64e	40.24b	116.68b
UFPI-278	9.86c	3.49b	0.43d	21.86a	64.11c	27.95c	9.42c	18.53d	98.22b
UFPI-280	9.34d	3.06e	0.46d	20.82a	66.49b	26.60c	5.82f	20.79d	100.94b
UFPI-282	9.66c	2.85f	0.41d	21.14a	65.59c	62.42a	9.31c	53.11a	66.58c
UFPI-468	9.04e	3.14d	0.16e	23.41a	63.94c	31.08c	9.91b	21.17d	102.01b
UFPI-491	9.13e	3.30m4c	0.88a	20.77a	66.46b	35.55c	8.39d	27.15c	71.88c
Mean Geral	9.68	3.21	0.50	20.76	65.85	39.82	8.52	31.31	83.77
CV %	2.83	8.13	20.26	6.50	2.17	16.67	8.27	21.49	19.90

[1]Means followed by the same letter, within the same column don’t significantly differ from each other, according to the Scott-Knott test ($p<0.05$).

Table 3. Mean values[1] for physical characters: seed length (L), width (W) and thickness (T) (in mm) and the 100-seed weight (100SW) (in g), and classification of the seeds in terms of their shape, profile and size for the 27 tested lima bean accessions. Teresina, Piauí, 2008

Accession	L	W	T	100SW	Shape	Profile	Size
UFPI-032	15.80b	10.30b	5.97c	54.00b	ELP	FL	N
UFPI-121	10.90e	8.88d	6.29b	38.49d	SPH	SFL	M
UFPI-134	14.00c	9.72c	5.45d	47.84c	ELP	FL	N
UFPI-173	15.40b	9.88c	6.06c	56.81b	ELP	FL	N
UFPI-177	13.70c	9.87c	5.94c	45.82c	SPH	FL	N
UFPI-189	14.10c	10.50b	6.39b	60.25b	SPH	FL	L
UFPI-216	12.00d	8.58d	5.45d	34.29d	SPH	FL	M
UFPI-220	9.66e	7.48e	5.94c	28.01d	SPH	SFL	S
UFPI-222	9.94e	7.68e	5.95c	29.41d	SPH	SFL	S
UFPI-228	11.10e	8.02e	5.42d	30.94d	SPH	SFL	M
UFPI-229	10.50e	8.16e	5.95c	33.11d	SPH	SFL	M
UFPI-230	10.50e	7.42e	5.58d	27.60d	SPH	SFL	S
UFPI-234	11.31e	8.21e	5.84d	34.47d	SPH	SFL	M
UFPI-237	10.60e	7.77e	5.61d	28.51d	SPH	SFL	S
UFPI-243	10.50e	7.94e	5.99c	30.69d	SPH	SFL	M
UFPI-251	12.50d	8.62d	5.33d	35.48d	ELP	FL	M
UFPI-264	12.40d	8.71d	5.37d	34.71d	SPH	FL	M
UFPI-271	13.40c	9.51c	5.91c	48.92c	SPH	FL	N
UFPI-272	15.50b	10.00b	6.21c	57.99b	ELP	FL	N
UFPI-274	15.70b	10.30b	6.12c	61.14b	SPH	FL	L
UFPI-275	14.90b	10.20b	6.10c	63.52b	ELP	FL	L
UFPI-276	18.50a	11.80a	6.91a	87.79a	ELP	FL	L

Accession	L	W	T	100SW	Shape	Profile	Size
UFPI-278	12.20d	9.25c	6.31b	47.40c	SPH	FL	N
UFPI-280	14.00c	10.10b	6.47b	56.46b	SPH	FL	N
UFPI-282	14.50c	10.30b	6.53b	55.57b	SPH	FL	N
UFPI-468	14.70b	10.10b	6.47b	47.71c	ELP	FL	N
UFPI-491	18.10a	10.60b	6.31b	85.17a	OKS	FL	L
Mean	13.23	9.27	5.99	46.74			
CV%	7.68	5.96	4.56	18.01			

[1]Means followed by the same letter within the same column do not significantly differ from each other', according to the Scott-Knott test ($p<0.05$).

Shape (SPH: spherical, ELP: elliptical, OKS: oblong/kidney-shaped, short); Profile (FL: flat, SFL: semi-flat); Size (P: small, M: medium, N: normal and L: large).

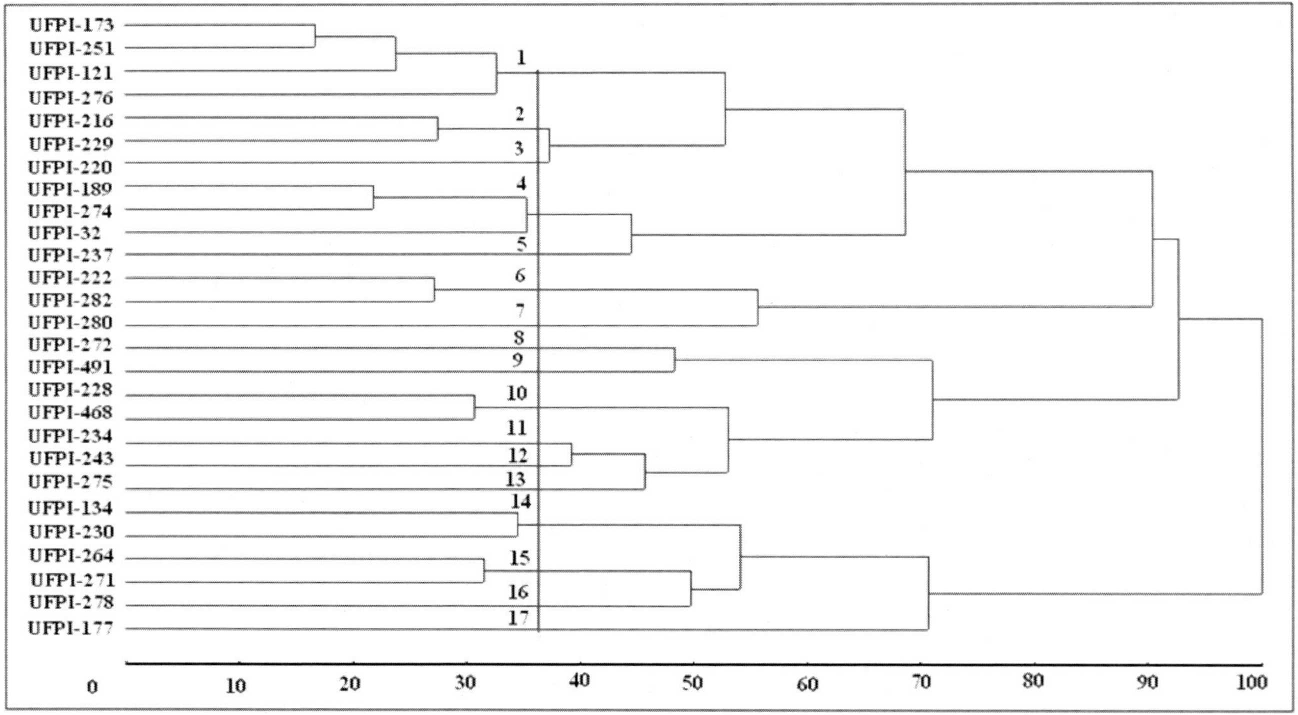

Figure 1. Dissimilarity dendrogram obtained using the UPGMA method, based on nine chemical and four physical characters from 27 lima bean genotypes. Teresina, Piauí, 2008.

Group 3 included four accessions exhibiting a high protein content (UFPI-189 and UFPI-468), a high non-nitrogen extract content (UFPI-177) and elliptical shape (UFPI-272 and UFPI-468). Group 4 included two accessions, which exhibited a high crude protein (UFPI-121 and UFPI-280) and neutral detergent fibre and hemicellulose contents (UFPI-121).

Group 5 included two accessions with a high crude protein (UFPI-134) and non-nitrogen extract content (UFPI-173), both of which exhibited a low hydrogen cyanide concentration and seeds with an elliptical shape. Group 6, also comprised of two accessions, exhibited high ether extract (UFPI-491) and crude protein contents (UFPI-276 and UFPI-491), a low hydrogen cyanide concentration (UFPI-491) and a high mean 100-seed weight (UFPI-276 and UFPI-491).

Accessions UFPI-32 (group 7) and UFPI-271 (group 8) presented high neutral detergent fibre and hemicellulose contents, respectively, and flat seeds with a normal size. Accession UFPI-220 from group 9 exhibited a high protein content, and accession UFPI-282 from group 10 was characterised by high crude protein and hemicellulose contents. It should be noted that the genotypes from these last two groups exhibited low hydrogen cyanide concentrations.

The performance of the tested accessions (Tables 2 and 3), the obtained measures of dissimilarity, and the groups of accessions established according to the Tocher method indicate that crossings of UFPI-491 with UFPI-282, UFPI-121, UFPI-220 and UFPI-229 represent a good option for obtaining segregating populations with high protein and fibre contents and a low hydrogen cyanide concentration. Ribeiro et al. (2005) found the Tocher method to be effective for the identification of promising progenitors for controlled hybridisation programs aimed at developing germplasms with above average protein and fibre contents. Mahalanobis's generalised distance can be used for the detection of genetic diversity between superior bean genotypes in breeding programs (Elias et al., 2007). However, gene control should also be considered because the accuracy of the predictions depends on the existence of dominance effects of genes involved in the expression of the characters of interest.

UFPI-275 and UFPI-274, which belonged to group 2, originated from the state of Paraíba, whereas UFPI-278 and UFPI-276, with the same origin, were placed in groups 1 and 6, respectively (Table 1). Genotypes UFPI-189, UFPI-121, UFPI-491 and UFPI-32, originating from Piauí, were placed in groups 3, 4, 6 and 7, respectively, while accessions UFPI-468, UFPI-280 and UFPI-282, collected in Maranhão, were placed in groups 3, 4 and 10 and UFPI-177 and UFPI-134, originating from Ceará, in groups 3 and 5.

This indicates that there was no relationship between genetic diversity, based on the examined physical and chemical characters of the seeds, and geographical origin, a finding that is in accord with other studies performed in climbing snap bean (Abreu et al., 2004] and lima bean (Guimarães et al., 2007).

The UPGMA hierarchical method clustered the 27 lima bean genotypes into 17 divergence groups, nine of which were formed by single genotypes. This shows that the UPGMA method is more restrictive than the Tocher method regarding the clustering of genotypes into groups (Figure 1). Values above the overall mean were observed for moisture content in groups 1, 7, 9, 10 and 11 (Tables 2 and 3); ash content in groups 1, 5, 9, 11, 13, 14, 15 and 17; ether extract content in groups 4, 5, 9, 11, 13, 14 and 16; crude protein content in groups 2, 6, 7, 9, 10, 11, 12, 13, 15 and 17; non-nitrogen extract content in groups 1, 3, 4, 5, 7, 8, 9, 10 and 16; neutral detergent fibre content in groups 1, 4, 6, 8, 11 and 12; acid detergent fibre content in groups 3, 4, 6, 8, 10, 11 and 15; hemicellulose content in groups 1, 4, 6, 8, 12 and 15; and the hydrogen cyanide concentration in groups 2, 4, 7, 8, 10, 11, 12, 15, 16 and 17.

The relative contribution of each character to the genetic divergence between the accessions was estimated according to Singh (1981). Insoluble (48.62%) and soluble fibre (46.03%) were the characters that contributed most to the total divergence (94.65%) between the 27 tested lima bean accessions. Seed thickness, 100-seed weight, the hydrogen cyanide concentration, seed width, and the non-nitrogen extract and ether extract contents contributed little to the observed divergence.

However, removing these characters from the analysis altered the original formation of groups, indicating that, despite their low contributions, these characters were important for establishing the divergence groups according to the method proposed by Oliveira et al. (2004).

Using the same method, Abreu et al. (2004) identified eight characters that contributed nothing to the observed genetic divergence, as their removal from the analysis did not change the original group composition. However, removing the 100-seed weight, which exhibited a low contribution, caused a change in the groups, and this character was therefore considered important regarding the divergence between accessions of common bean (*P. vulgaris*). According to Coelho et al. (2007), who also performed studies on bean accessions, yield-related characters such as the 100-seed weight are more discriminating in more uniform populations of landrace genotypes.

Variability in the physical and chemical characters of seeds was observed for the tested lima bean accessions from the Active Germplasm Bank of Piauí

Federal University. The insoluble and soluble fibre contents were the characters that contributed most to the total divergence between accessions, thus allowing the selection of divergent progenitors to obtain segregating populations.

ACKNOWLEDGMENTS

The authors wish to thank the National Council for Scientific and Technological Development (Conselho Nacional de Desenvolvimento Científico e Tecnológico - CNPq) for the master's scholarship granted to the first author and for financing the *P. lunatus* research project.

REFERENCES

Abreu, F.B., Leal, N.R., Rodrigues, R., Amaral JR., A.T, Silva, D.J.H., (2004). Divergência genética entre acessos de feijão-de-vagem de crescimento indeterminado. *Horticultura Brasileira* 22, 547–552.

Association of Official Analytical Chemists (AOAC) International, (2005). *Official methods of analysis of AOAC International*, 18th ed., Gaithersburg: AOAC International, pp 1- 2, pp 53-56.

Azevedo, J. de N., Franco, L.J.D., Araújo, R.O.C., (2003). *Composição química de sete variedades de feijão-fava* . Teresina: Embrapa Meio Norte. pp 4. (Comunicado Técnico, 152).

Carvalho, L.P. de, Lanza, M.A., Fallieri, J., Santos, J.W. dos, (2003). Análise da diversidade genética entre acessos de banco ativo de germoplasma de algodão. *Pesquisa Agropecuária Brasileira* 38,1149-1155.

Coelho, C.M.M., Coimbra, J.L.M.,Souza, C.A., Bogo, A., Guidolin, A. F., (2007). Diversidade genética em acessos de feijão (*Phaseolus vulgaris* L.). *Ciência Rural* 37, 1241-1247.

Cruz, C.D., Regazzi, A.J., Carneiro, P.C.S. (2004). *Modelos biométricos aplicados ao melhoramento genético*. 3rd. ed. Viçosa: UFV, pp 480.

Cruz, C.D. (2007). *Programa Genes* – Windows version. Viçosa: UFV.

Elias, T.H., Goncalves-Vidigal, M.C., Gonela, A. Vogt, G.A., (2007). Variabilidade genética em germoplasma tradicional de feijão-preto em Santa Catarina. *Pesquisa Agropecuária Brasileira* 42, 1443-1449.

Guimarães, W.N.; Martins, L.S.; Silva, E.F.; Ferraz, G.M.G.; Oliveira, F.J., (2007). Caracterização morfológica e molecular de acessos de feijão-fava (*Phaseolus lunatus* L.) *Revista Brasileira de Engenharia Agrícola and Ambiental* 11, 37-45.

Karasawa, M.; Rodrigues, R.; Sudré, C.P.; Silva, M.P.; Riva, E.M.; Amaral Júnior, A.T., (2005). Aplicação de métodos de agrupamento na quantificação da divergência genética entre acessos de tomateiro. *Horticultura Brasileira* 23,1000-1005.

Lymman, J.M., (1983). Adaptation studies on lima bean accessions in Colombia. *Journal-American Society for Horticultural Science* 108, 369-373.

Oliveira, A.C.B., Sediyama, M.A.N., Pedrosa, M.W., Garcia, N.C.P., Garcia, S.L.R., (2004). Divergência genética e descarte de variáveis em alface cultivada sob sistema hidropônico. *Acta Scientiarum* 26, 211-217.

Oliveira, F.J., Anunciação Filho, C.J. da, Bastos, G.Q., Reis, O.V. dos, (2003). Divergência genética entre cultivares de caupi [Genetic divergence among cultivars of cowpea]. *Pesquisa Agropecuária Brasileira* 38, 605-611.

Pérez-Herrera, P., G. Esquivel-Esquivel, R. Rosales-Serna, y A. Acosta-Gallegos., (2002). Caracterización física, culinaria y nutricional de frijol del altiplano sub-húmedo de México. *Archivos Latinoamericanos de Nutrición* 52,172-180.

Rachie, K.O., Song, L., Lyman, J., (1980) Lima bean (*Phaseolus lunatus*) and potential in the tropics. In: Summerfield, R. J.; Bunting, A. H. (Eds.). *Advances in legume science.* Kew: Royal Botanic Garden, pp 375-381.

Rao, R.C., (1952). *Advanced statistical methods in biometric research.* New York: J. Wiley, pp 390.

Ribeiro, H.J.S.S., Prudencio-Ferreira, S.H., Miyagui, D. T., (2005). Propriedades físicas e químicas de feijão comum preto; cultivar IAPAR 44; para envelhecimento acelerado. *Ciência and Tecnologia de Alimentos* 25, 165-169.

Santos, E.R., Santos, E.R., Barros, H.B., Ferraz, E. de C., Capone, A., Santos, A.F. dos, Fidelis, R.R., (2011) et al., (2011). Divergência entre genótipos de soja, cultivados em várzea irrigada. *Revista Ceres* 58, 755-764.

Silva, D.J., (1990) Análise de alimentos (métodos químicos and biológicos). Viçosa: UFV, pp 165.

Singh, D., (1981). The relative importance of characters affecting genetic divergence. *The Indian Journal of Genetic and Plant Breeding* 41, 237-245.

Vilhordo, B.W., Mikusinski, O.M.F., Burin, M.E., Gandolfi, V.H., (1996). Morfologia. In: Araújo, R.S., Rava, C.A.; Stone, L.F.; Zimmermann, M.J.O. (Coord.). *Cultura do feijoeiro comum no Brasil.* Piracicaba: Potafos, pp 71-99.

In: Phaseolus Lunatus
Editors: A. S. Ferreira de Araujo et al.
ISBN: 978-1-63482-494-1

Chapter 3

RECENT ADVANCES IN THE STUDY OF THE EVOLUTION OF LIMA BEAN (*PHASEOLUS LUNATUS* L.) IN MEXICO

Jaime Martínez-Castillo[#,1]**,**
Rubén Humberto Andueza-Noh[†,2]
and María Isabel Chacón Sánchez[‡,3]

[1]Centro de Investigación Científica de Yucatán, Yucatán, México
[2]División de Estudios de Posgrado e Investigación, Instituto Tecnológico de Conkal-Catedras-Conacyt Avenida Tecnológico, Conkal, Yucatán, México
[3]Universidad Nacional de Colombia - Bogotá - Facultad de Ciencias Agrarias - Departamento de Agronomía- Bogotá, D.C., Colombia

ABSTRACT

The origin of agriculture and the crop domestication are two of the most important processes in human history. Mexico is considered one of the main centers of origin of agriculture and domestication of plants in the world. Evidence of domestication of several species of worldwide

---

Email: jmartinez@cicy.mx
† Email: r_andueza81@hotmail.com
‡ Email: michacons@unal.edu.co

importance exists there, for example the *Phaseolus* beans. Lima bean (*P. lunatus*) is one of the *Phaseolus* species domesticated in Mexico. In this country, it is a subsistence crop that is cultivated mainly by ethnic groups as part of their traditional agricultural systems such as the milpa. Alongside with Lima bean landraces, in Mexico there are wild populations of this species. Regional studies have indicated the existence of high levels of genetic diversity within this crop and its wild populations and also have highlighted the importance of Mexico for the conservation of this species. Recently, molecular studies that have included new collections of wild Lima bean throughout Mexico have confirmed the high levels of genetic diversity in this species and suggested the existence of three wild Mesoamerican gene pools. Also, these studies suggest a scenario of multiple domestication events for small-seeded Mesoamerican Lima bean landraces in two putative domestication areas, one of them in Mexico (in the Nayarit-Jalisco or Guerrero-Oaxaca areas) and another one in a region between Guatemala and Costa Rica. Although there have been significant recent advances in our knowledge about the evolution of Lima bean in Mexico, it is necessary to dig deeper into this important topic. New collections of wild and domesticated material in specific sites of Mexico and the application of new techniques of massive sequencing of DNA will be of great help into this matter.

INTRODUCTION

Crop domestication, along with the development of agriculture, is considered one of the most important processes in the world, because it has allowed the development and maintenance of human society in the last 11,000 years, causing the shift in the lifestyle from an economy based on hunting and gathering of wild plants and animals to an economy based on the production of domesticated species, being a mayor transition in human history. Some hypotheses suggest that both processes were initiated between 11,000 and 5,000 years ago in six different regions around the world, primarily in tropical and subtropical areas with high biological and cultural diversity including China, Southeast Asia, Africa, the Near East, the Southern Andes and Mesoamerica (Harlan, 1992; Smartt and Simmonds, 1995; Piperno and Pearsall, 1998; Smith, 2005; Gepts, 2008).

Mesoamerica, cultural region geographically placed between the tropical limits of Mexico in the north, and the Central Valley of Costa Rica in the south (Kirchof, 1943), is considered one of the world's primary centers of crop domestication (Harlan 1995). In this wide territory, a large diversity of plants can be found as well as a great diversity of cultural groups that were involved in crop domestication and the origin of agriculture. Based on the diversity of plant species that have been domesticated in Mesoamerica, this region is recognized as one of the areas of the New World where agriculture was first practiced (MacNeish, 1967; Harlan, 1975; Flannery, 1986). In Mexico, the major Mesoamerican's region and center of origin of agriculture and cradle of major Mesoamerican crops (Zizumbo-Villarreal and Colunga, 2010), two areas of origin have been proposed: a) the intermountain valleys of central Mexico (Tehuacán-Puebla; MacNeish and Eubanks, 2000; Eubanks, 2002) and b) the low areas of the southwest in the basin of the Balsas river (Doebley et al., 2006, Piperno et al., 2007; Machuca-Chavéz et al., 2010; Zizumbo-Villarreal and Colunga-GarcíaMarín, 2010). In these areas, crops like squash (*Cucurbita pepo* L; Sanjur et al., 2002), maize (*Zea mays*, Fukunaga et al., 2005; Buckler et al., 2006; Matsuoka et al., 2002), chilies (*Capsicum* spp.; Pickersgill 1989; Aguilar-Meléndez et al. 2009) and beans (*Phaseolus* spp., Gepts, 1990; Chacón et al., 2005; Motta-Aldana et al., 2010; Serrano-Serrano et al., 2012; Bitochi et al., 2013; Andueza-Noh et al., 2013; Schmutz et al., 2014; Gaut, 2014) were domesticated.

In this paper, we provide a review of the state of the art and recent advances on the domestication history of Lima bean (*Phaseolus lunatus* L.) in the Americas, with special reference to Mexico. First, we briefly highlight the importance of beans and Lima bean in Mexico, and then we focus on the evolutionary history of Lima bean as a wild species, specifically its origin in the northern Andes and its current genetic structure. We also address the history of its domestication in the Americas, providing evidence for at least two areas of domestication for small-seeded landraces, one of them in Mexico. We conclude our review by pointing some unanswered questions and giving some insights in what should be the focus of future research in this species.

BEAN DOMESTICATION (*PHASEOLUS* SPP.) IN MEXICO

The genus *Phaseolus* is of American origin. It contains more than 50 wild species, but only five of them have been reported as domesticated, namely *Phaseolus vulgaris* L. (common bean), *Phaseolus coccineus* L. (scarlet runner bean), *Phaseolus dumosus* Macfad. (year bean), *Phaseolus acutifolius* A. Gray (tepary bean) and *Phaseolus lunatus* L. (Lima bean) (Freytag and Debouck, 2002). These five species belong to two separate monophyletic species clades, the Vulgaris group to which the first four species belong along with other wild species from Mesoamerica, and the Lunatus group to which the latter species belong along with other wild species mainly from South America (Delgado-Salinas et al., 1999).

In Mexico, the domestication of *Phaseolus* beans has been a widely studied topic because they have played an important role in the food culture of different ethnic groups in this country.

Figure 1. Examples of *Phaseolus* beans cultivated in Mexico. A, *P. acutifolius*; B, *P. coccineus*; C, *P. vulgaris*; D, *P. lunatus*. The first seed at the left in each case is the wild form. For A, C and D, there are evidence about their domestication in Mexico. Photograph taken by J. Martínez-Castillo.

The radiocarbon dating results of Kaplan (1965) suggested that beans are one of the oldest crops in the New World with a date of domestication as early as 7,000 years ago for the common bean in Mexico. Later, Kaplan and Lynch (1999) provided revised dates of archaeobotanical records by accelerator mass spectrometry and indicated a much later domestication date for beans in Mexico between 2,500 and 1,300 years ago. Beans, together with maize, squash and chilies, apart from being an old crop have also been a dominant staple in Mexico because of its high nutritional value (Zizumbo-Villarreal and Colunga-GarcíaMarín, 2010), which is reflected in the extreme diversity found in cultivation methods, uses, and the range of environments to which they have been adapted (Broughton et al., 2003; Lopez-Soto et al., 2005).

In the last years, several studies have indicated the importance of Mexico as a center of domestication of at least three *Phaseolus* bean species (see Figure 1). Molecular analyses using markers as phaseolin (the main seed storage protein), microsatellite markers and sequence data point to Mexico as a center of domestication for common bean, in particular the area that includes the lower basin of the Lerma River, the Rio Grande de Santiago, Rio Verde in Jalisco and the Mascota-Ameca river, near the Chapala-Sayula-Zacoalco lake system (Kwak et al., 2009; Nanni et al., 2011; Bitocchi et al., 2012). Results using AFLP markers have indicated one of two Mexican states, Sinaloa or Jalisco, as a domestication area of tepary bean (Muñoz et al., 2006). In Lima bean, recent studies using noncoding regions of the chloroplast DNA (cpDNA) and internal transcribed spacer (ITS) polymorphisms of ribosomal DNA (rDNA) (Motta-Aldana et al., 2010; Serrano-Serrano et al., 2010, 2012; Andueza-Noh et al., 2013) have indicated western central Mexico as a domestication area of Lima bean, as discussed below.

IMPORTANCE OF LIMA BEAN IN MEXICO

According to Baudet (1977), Lima bean is composed of wild (*P. lunatus* var. *silvester*) and domesticated (*P. lunatus* var. *lunatus*) populations. In Mexico, Lima bean is cultivated by diverse ethnic groups who have named it in different ways: "Ib or Ibes" by Mayan groups from the Yucatan Peninsula, "Pathasetes" by Zoques from Chiapas, and "Pataxtles" by Nahuatls from Veracruz, among others (Ballesteros, 1999). In Mexico, Lima bean is not cultivated intensively as in some others countries (e.g., USA, Peru, Madagascar) (Brink and Belay, 2006) and its importance is mainly as a crop of subsistence, being part of traditional cropping systems such as the milpa or home-gardens. Among the main Mexican states where Lima bean is cultivated are Veracruz, Guerrero, Michoacán, Oaxaca, Chiapas, Campeche, Quintana Roo and Yucatan (Ballesteros, 1999).

Figure 2. Lima bean plants growing on dead corn plants in a Mayan milpa from Yucatan state, Mexico. Photograph taken by J. Martínez-Castillo.

An example of the importance of Lima bean as a subsistence crop is the traditional Mayan agriculture from Yucatan Peninsula, Mexico. This region comprises the states of Campeche, Quintana Roo and Yucatan. There, Lima bean landraces are part of the milpa, namely the Mesoamerican farming system based on the slash and burn of the vegetation (Figure 2). In this region, Lima bean is a secondary crop associated with maize, squash, and other species. In 1999, Ballesteros undertook a botanical exploration of Mexico and recorded the highest diversity of cultivated variants of Lima bean (in terms of seed characters) in the Yucatan Peninsula, included in 25 local names. Martínez-Castillo et al. (2004) confirmed the high richness of landraces present in the Yucatan Peninsula and showed the great importance of this crop for the traditional Mayan farmers: Lima bean landraces are the fourth most important crop, only after maize, common bean and squash. Also, Martínez-Castillo et al. (2006, 2007, 2008) using molecular markers, showed that this region has high levels of genetic diversity, not only in the landraces, but also in the wild populations of this species. Recently, Martínez-Castillo et al. (2014) showed the existence of wild-crop genetic introgression in *P. lunatus* and the important role played by the Mayan farmers into this process. All these studies indicate the importance of this Mexican region for the conservation of this species.

EVOLUTIONARY HISTORY OF WILD LIMA BEAN

Wild populations that are ancestors to crops are of special importance because they contain alleles that at some point may become useful for future or current breeding programs (Hajjar and Hodgkin, 2007). In order to exploit better the genetic diversity locked in wild relatives, it is necessary to understand how this diversity is organized along the geographic range where wild populations occur, and how different evolutionary processes have shaped current patterns of genetic structure in these populations (Loveless and Hamrick, 1984). It is widely accepted that all this knowledge is key not only to design sampling and conservation strategies (Hamrick, 1983; Hedrick, 2001), but also to understand domestication history of crops (Kwak and Gepts, 2009; Angioi et al., 2010).

Wild Lima bean populations, as well as domesticated, are widely distributed in the Americas, in a geographical range that runs from northern Mexico to northern Argentina, in regions of low to mid elevations and seasonal climates (Freytag and Debouck, 2002) (Figure 3). Along this wide geographical range, two main gene pools have been recognized, the small-seeded Mesoamerican gene pool and the large-seeded Andean gene pool, on the basis of morphological, biochemical and molecular evidence (Gutiérrez-Salgado et al., 1995; Lioi et al., 1998; Caicedo et al., 1999; Fofana et al., 2001; Lioi and Galasso, 2002). The distribution of the Mesoamerican gene pool starts in the state of Sonora in northern Mexico and continues along the pacific slope of the Mexican mountains and along the plains of the Gulf of Mexico and the Peninsula of Yucatan, entering into Guatemala, Belize, El Salvador, Honduras, Nicaragua, Costa Rica, Panama and in some few Caribbean islands (Freytag and Debouck, 2002). According to herbarium records and germplasm collections held at the International Center for Tropical Agriculture – CIAT (Debouck, 2014), within Mexico wild Lima bean populations have been found in 19 states, whose names in alphabetic order are Baja California Sur, Campeche, Chiapas, Colima, Guerrero, Jalisco, México, Michoacán, Morelos, Nayarit, Oaxaca, Puebla, Quintana Roo, Sinaloa, Sonora, Tabasco, Tamaulipas, Veracruz and Yucatán. The Mesoamerican gene pool is also found in South America, from Colombia to Argentina, on the eastern slope of the Andes. On the other hand, the Andean gene pool is restricted to the western slope of the Andes, in an area between southern Ecuador and northern Peru.

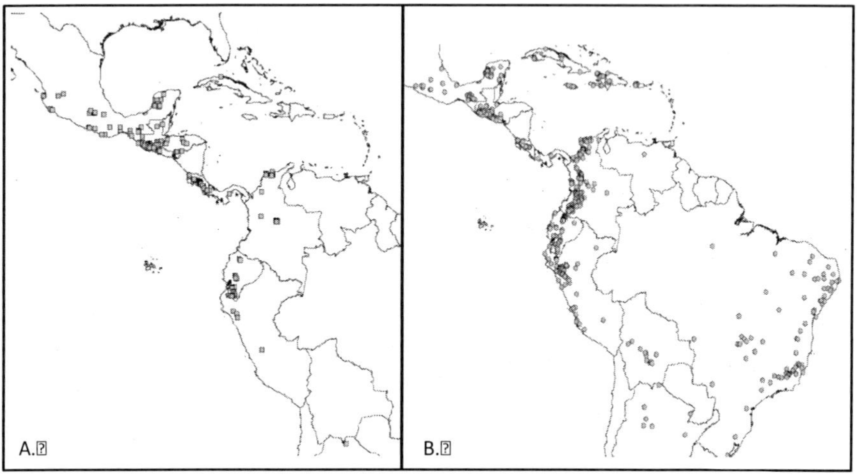

Figure 3. Geographic distribution of wild (A.) and domesticated (B.) accessions of Lima bean in the Americas according to the germplasm accessions held at CIAT (International Center for Tropical Agriculture, www.ciat.cgiar.org).

How wild Lima bean has reached its current distribution is an interesting biogeographical question because Mesoamerica and South America have been connected through the Isthmus of Panama probably since the Middle Miocene (about 12 million years ago, mya) or younger times (Montes et al., 2012). Therefore, several studies have focused on the evolutionary history of wild Lima beans using biochemical and molecular evidence. First, Fofana et al. (1999) carried out a phylogenetic analysis on the basis of chloroplast DNA (cpDNA) polymorphisms from the two gene pools of Lima bean and other allied species from Mesoamerica and South America, confirming *P. lunatus* as monophyletic and proposing an Andean origin for this species on the basis of a closer genetic relationship of *P. lunatus* with Andean allied species (*P. augusti*, *P. pachyrrhizoides* and *P. bolivianus*). Later, Sparvoli et al. (2001) also suggested an Andean origin of *P. lunatus* according to the polymorphisms found in three genes of the Lectin family of proteins.

These results of an Andean origin for the species allowed hypothesizing that, after its origin, wild *P. lunatus* would have colonized Mesoamerica by crossing the Isthmus of Panama, thus reaching its current distribution. Therefore, further studies were performed to assess the hypothesis of an Andean origin of the species and to understand the processes underlying its wide distribution. For this, Serrano-Serrano et al. (2010) sequenced the Internal Transcribed Spacer (ITS) of the ribosomal DNA and two cpDNA intergenic spacers (*trn*L-*trn*F and *atp*B-*rbc*L) in a sample of 59 wild Lima bean accessions and allied species. The results confirmed an Andean origin of wild Lima bean in Pleistocene times (at about 1 mya), namely at an age that is posterior to the formation of the Isthmus of Panama (according to Montes et al., 2012), making plausible the hypothesis of a colonization of Mesoamerica from the Andes through the Isthmus of Panama.

In their study, Serrano-Serrano et al. (2010) also proposed a more complex structuring of the genetic diversity of wild Lima bean into three gene pools, two Mesoamerican gene pools called MI and MII, and the Andean one. While gene pool MII is widely distributed in South America and Mesoamerica, gene pool MI is restricted to Mexico. Gene pool MI occurs along the pacific slope of the Mexican mountains, from the State of Sinaloa to the state of Oaxaca and it is found mainly to the northwest of the Isthmus of Tehuantepec (Figure 4).

Figure 4. Geographic distribution of gene pools MI (closed circles) and MII (gray squares) in wild *Phaseolus lunatus* from Mexico and neighboring countries, and putative domestication areas of Lima bean in Mexico (dotted lines).

On the other hand, gene pool MII is found in the plains of the Gulf of Mexico, in the states of Tamaulipas, Veracruz and Tabasco, in the Peninsula of Yucatan, the state of Chiapas and is also found in Central America from Guatemala to Costa Rica. In South America, gene pool MII is found on the eastern slope of the Andes and in lands of low and mid elevation in Venezuela, Colombia, Ecuador, Peru and Argentina. On the other hand, the Andean gene pool is restricted to the western slope of the Andes of southern Ecuador and northern Peru. These distribution patterns suggest that the Andean orogeny and the emergence of the Isthmus of Panama had an influence in the current distribution of these gene pools.

The fact that both Mesoamerican gene pools, MI and MII, occur in Mexico and with mostly non-overlapping ranges (see Figure 4), makes this country an area of interest for understanding the genetic structure of this species and its relation with geography. This is what motivated Martínez-Castillo et al. (2014) to carry out a study of wild Lima bean from Mexico with a set of ten microsatellite markers (SSR) analyzed in a sample of 67 wild populations that cover the known distribution of the species in this country. The results support the existence of two gene pools of wild Lima bean in Mexico, MI and MII, and suggest a possible split of MI into two groups called MIa and MIb, although these results need further confirmation. While gene pool MIa was mainly observed in the states of Sinaloa, Nayarit, Jalisco, Guanajuato, Colima, Michoacán, Guerrero, Oaxaca, Morelos and Puebla; gene pool MIb was only found in Nayarit, Jalisco, Colima, and Morelos. MII was observed in the plains of the Gulf of Mexico in the states of Tabasco and Veracruz, southern Oaxaca, in Chiapas and in the Peninsula of Yucatan. Wild *P. lunatus* from Mexico contained relatively high levels of genetic diversity (expected heterozygosity, H_E= 0.61), and the gene pools MI (H_E=0.54) and MII (H_E=0.52) showed similar levels of genetic diversity. In summary, gene pools MI and MII in Mexico are mostly non-overlapping in their geographic distribution, are highly differentiated and contain comparable levels of genetic diversity; therefore these two gene pools should be considered as separate conservation units.

DOMESTICATION HISTORY OF LIMA BEAN

The origin of domesticated plants is an issue that has interested to scholars since long time ago. To understand the evolutionary history of crop plants, one of the main aspects to study is the biological origin, that is, the wild ancestor from which the crop was domesticated, and a second aspect is the geographic origin of the crop that is the geographic area where humans took into cultivation and domesticated the wild ancestors. This last aspect is of particular relevance for widely distributed wild species, as the Lima bean, since several human groups could have domesticated the crop independently in different geographical regions. This is what happened with Lima bean, where evidence today strongly supports that the small-seeded Mesoamerican landraces were domesticated independently from the large-seeded Andean landraces.

One of the earliest hypotheses on the origin of Lima bean landraces and proposed by Mackie (1943) indicated that all the landraces, small and large-seeded, had a single origin. The author, by examining samples of wild and cultivated Lima beans received from different sources, proposed a single origin of the crop in Guatemala. From this place of origin, the author proposed three migration routes throughout the Americas, two of them for the small-seeded landraces, namely the Hopi Branch and the Carib or West Indies Branch, and the Inca Branch for the large-seeded landraces. Through the Hopi Branch, Lima bean landraces would have migrated to the north as far as the United States, and through the Carib Branch they would have migrated to Yucatán, the Caribbean, Venezuela and Brazil. Through the Inca Branch, the large-seeded landraces would have reached Central America, Colombia, Ecuador and Peru. These hypotheses were based on the analyses of a limited sample of wild Lima beans recovered from the Guatemala area and landraces from different geographic regions.

Although a large diversity of seed morphotypes is found among Lima bean landraces (Martínez-Castillo et al., 2004), two main groups have been recognized, the small-seeded landraces that are mainly distributed in Mesoamerica, the Caribbean and South America, and the large-seeded landraces that are mainly found in the Andes of Colombia, Ecuador, Peru and Bolivia (Debouck et al., 1989). As we have seen above, in wild Lima bean two gene pools have been recognized, the Mesoamerican one of small seeds and the Andean one of large seeds. In addition, and as we should see below, there is also a wide body of evidence that shows that small-seeded landraces are genetically related to wild populations from the Mesoamerican gene pool, and on the other hand, that large-seeded landraces are genetically related to wild populations from the Andean gene pool, thus supporting the hypothesis of at least two domestication events, one in the Mesoamerican gene pool and one in the Andean gene pool.

Gutiérrez-Salgado et al. (1995) studied electrophoretic patterns of seed proteins in a sample of 84 wild, 6 weedy and 426 cultivated accessions of Lima beans. They found that large-seeded landraces could have been domesticated from wild forms in the Andean gene pool distributed in western South America. Also, they found that the electrophoretic pattern that was most widespread among the small-seeded landraces was also common among wild forms of the wild Mesoamerican gene pool and thus could not determine their place of origin. In a later study, Maquet et al. (1997) on the basis of electrophoretic patterns of allozyme alleles also found that large-seeded landraces were domesticated from wild forms in the Andean gene pool in the western valleys of Ecuador and northern Peru, and although wild and domesticated small-seeded types were grouped together within the same gene pool, the authors could not establish the place of origin of the Mesoamerican landraces. Nienhuis et al. (1995) and Fofana et al. (1997) further confirmed the existence of two main gene pools, the Mesoamerican and the Andean one, each consisting of wild and landrace accessions, on the basis of RAPD markers, thus supporting the hypothesis of at least two domestication events. Further support for the existence of these two major gene pools and two independent domestications was given by Caicedo et al. (1999) on the basis of AFLP markers and by Fofana et al. (2001) on the basis of cpDNA polymorphisms, but without addressing the issue of the geographic origin of the Mesoamerican landraces.

More recent research have dealt with the study of the geographic origin of the Mesoamerican landraces of Lima bean (Motta-Aldana et al., 2010; Serrano-Serrano et al., 2012; Andueza-Noh et al., 2013) on the basis of DNA sequence polymorphisms of the ITS region of the ribosomal DNA and of two intergenic spacers of the cpDNA (*trn*L-*trn*F and *atp*B-*rbc*L). As we have seen above, Serrano-Serrano et al. (2010) proposed, on the basis of these three genomic regions, the existence of three gene pools in wild Lima beans, two Mesoamerican (MI and MII) with mostly non-overlapping ranges, and the Andean one. Motta-Aldana et al. (2010) in analyzing wild and cultivated forms found that most of the small-seeded landraces analyzed grouped into gene pool MI, thus suggesting a domestication event in the area to the north and north-west of the Isthmus of Tehuantepec in Mexico, namely within the distribution area of the MI gene pool (Figure 4).

Later, Serrano-Serrano et al. (2012) and Andueza-Noh et al. (2013) using these same three genomic regions in a larger sample of wild and landrace accessions from Mexico proposed not only one but two possible areas of domestication for the Mesoamerican landraces, one of them from the MI gene pool in the states of Nayarit-Jalisco or Guerrero-Oaxaca in Mexico (Figure 4), and a second one from the gene pool MII in an area between Guatemala and Costa Rica. These two studies consequently suggest the possibility of multiple origins for the small-seeded Mesoamerican landraces in two separate geographic areas, although further confirmation is needed with a more complete genomic sampling. An interesting result of these studies is that currently the largest diversity of cultivated forms within Mexico is in the

Peninsula of Yucatan (Ballesteros, 1999; Martínez-Castillo et al., 2004), a region that is outside the putative domestication areas, that harbors wild populations with high levels of genetic diversity and that is home of traditional agricultural societies.

CONCLUSION

Mexico has been identified as an important domestication center for many important crops, among these, beans (*Phaseolus* spp.). In the particular case of Lima bean, central-western Mexico has been identified as one of its areas of domestication (Motta-Aldana et al., 2010; Serrano-Serrano et al., 2012; Andueza-Noh et al., 2013). The location of the areas of domestication is the first step in understanding the process of adaptation of wild populations to cultivated conditions and also the genetic basis of the domestication syndrome.

Martínez-Castillo et al. (2008, 2012) also showed that Mexico is an important area of genetic diversity for Lima bean landraces, especially, the Yucatan Peninsula region. These same authors showed the high risk of genetic erosion existing for these landraces and recommended this region as an important conservation area for this crop. Yucatan Peninsula is also considered an important area of genetic diversity for wild populations of Lima bean (Martínez-Castillo et al., 2006, 2007). Recently, Martínez-Castillo et al. (2014) confirmed this high diversity through Mexico and showed that the organization of this diversity could be more complex than previously thought, with the existence of MII, MIa, and MIb gene pools. These authors indicated that the large genetic differentiation found among the three gene pools suggests that these should be considered as separate conservation units.

Although recent studies (Motta-Aldana et al., 2010; Serrano-Serrano et al., 2012; Andueza-Noh et al., 2013, Martínez-Castillo et al. 2014) have generated important advances in our knowledge on the evolution of Lima bean in Mexico, there are still unanswered questions where future research should focus. First of all, it is necessary to study in more deep the genetic structure of wild Lima bean in Mexico, because the presence of the two Mesoamerican gene pools (MI and MII) and the wide range of ecological conditions exploited by Lima beans make this a region of interest to understand, at a finer geographical scale, the processes that have given rise to these two gene pools, including adaptation to climatic conditions. Second, the scenario of multiple domestication events for the Mesoamerican landraces is extremely interesting, especially because it may partially explain the origin of the large amount of genetic diversity observed in landraces in the field, as it is the case for the Peninsula of Yucatan. In this same order of ideas, it is important to further analyze the role of wild-crop introgression in sustaining this diversity. Although we can consider that the geographical sampling of wild populations of *P. lunatus* from Mexico may be close to complete, new collections of wild and domesticated germplasm in specific sites of Mexico could be required. Also, new techniques of massive sequencing of DNA can be of great help to solve these questions by enabling a much larger sampling of the Lima bean genome. In this regards, we can say that the sequencing of the genome of Lima bean is urgently needed to advance in the understanding of the genetic diversity in this species.

REFERENCES

Aguilar-Meléndez, A., Morell, L. P., Roose, L. M., and Seung-Chul, K. (2009). Genetic diversity and structure in semiwild and domesticated chiles (*Capsicum annuum*; Solanaceae) from Mexico. *American Journal of Botany,* 96: 1190-1202.

Andueza-Noh, R. H., Serrano-Serrano, M. L., Chacón Sánchez, M. I., Sánchez del Pino, I., Camacho-Pérez, L., Coello-Coello, J., Mijangos Cortes, J., Debouck, D. G., and Martínez-Castillo, J. (2013). *Multiple domestications of the Mesoamerican gene pool of Lima bean* (*Phaseolus lunatus* L.): evidence from chloroplast DNA sequences. Genetic Resources and Crop Evolution, 60: 1069-1086.

Ballesteros, P. G. (1999). Contribuciones al conocimiento del frijol lima (*Phaseolus lunatus* L.) en América tropical. Tesis de doctorado en ciencias C.P. Montecillo, Texcoco, Estado de México. 386p.

Baudoin, J. P., Rocha, O., Degreef, J., Maquet, A., and Guarino, L. (2004). Ecogeography, Demography, Diversity and Conservation of *Phaseolus lunatus* L. in the Central Valley of Costa Rica. Systematic and ecogeographic studies on crop genepools. Internacional Plant Genetic Resources Institute, Rome, Italy. 94 p.

Bitocchi, E., Nanni, L., Bellucci, E., Rossi, M., Giardini, A., Spagnoletti Zeuli, P., Logozzo, G., Stougaard, J., McClean, P., Attene, G., and Papa, R. (2012). Mesoamerican origin of the common bean (*Phaseolus vulgaris* L.) is revealed by sequence data. *Proceedings of the National Academy of Sciences,*109: E788–E796.

Brink, M., and Belay, G. (2006). Plant resources of tropical Africa 1. Cereals and pulses. Programme PROTA.

Brougthon, W. J., Hernández, G., Blair, M., Beebe, S., Gepts, P., and Vanderleyden, J. 2003. Beans (*Phaseolus* spp.): model food legumes. *Plant and Soil,* 252: 55-128.

Buckler, E. S., Goodman, N. M., Holstford, T. P., Doebley, J. F., and Sánchez, J. (2006). Phylogeography of the wild subspecies of Zea mayz. Maydica, 51: 123-134.

Caicedo, A. L., M. C. Duque, O. Toro, D. Debouck, and J. Tohme. (1999). AFLP fingerprinting of *Phaseolus lunatus* L. and related wild species from South *America. Crop Sci., 39*:1497–1507.

Chacón, M. I., Pickersgill, S. B., and Debouck, D. G. (2005). Domestication patterns in common bean (Phaseolus vulgaris L.) and the origin of the Mesoamerican and Andean cultivated races. *Theoretical and Applied Genetics,* 110: 432-444.

Debouck, D. G., Maquet, A., and Posso, C. E. (1989). Biochemical evidence for two different gene pools in Lima beans, *Phaseolus lunatus* L. *Annu. Rep. Bean Improv. Coop*. (USA) 32: 58–9.

Debouck, D. G. (1999). Diversity in *Phaseolus* species in relation to the common bean. In Common Bean Improvement in the 21st Century. Ed. S R Singh. pp. 25–52. Kluwer Academic Publisher, Dordrecht.

Debouck, D. G. (2014). Cahiers de phaseologie: Section Paniculati, p. 390. CIAT, Cali, Colombia.

http://isa.ciat.cgiar.org/urg/showunitfiles.do?type=unitfiles&category=unitfiles &by=&collection= [accessed in October 2014].

Delgado-Salinas, A., Turley, T., Richman, A., and Lavin, M. (1999). Phylogenetic analysis of the cultivated and wild species of *Phaseolus* (Fabaceae). *Systematic Botany*. 24:438–460.

Delgado-Salinas, A., Bibler, R., and Lavin, M. (2006). Phylogeny of the genus *Phaseolus* (Leguminosae): A recent diversification in an ancient landscape. *Systematic Botany*, 31:779–791.

Doebley, J. F., Gaut, B. S., and Smith, B. D. (2006). The molecular genetics of crop *domestication. Cell,* 127:1309-1321.

Eubanks, M. (2002). "An interdisciplinary perspective on the origin of maize". *Latin American Antiquity,* 12:91-98.

Flannery, K.V. (1986). Guila Naquitz, archaic foraging and early agriculture in Oaxaca Mexico. Academic Press. New York.

Fofana, B., Vekemans, X., du Jardin, P., and Baudoin, J. P. (1997). Genetic diversity in lima bean *(Phaseolus lunatus* L.) as revealed by RAPD markers. Euphytica, 95:157-165.

Fofana, B., Baudoin, J. P., Vekemans, X., Debouck, D. G., and du Jardin, P. (1999). Molecular evidence for an Andean origin and a secondary gene pool for the Lima vean (*Phaseolus lunatus* L.) using chloroplast DNA. *Theor. Appl. Genet.* 98, 202–212.

Fofana, B., du Jardin P., and Baudoin J. P. (2001). Genetic diversity in the Lima bean (*Phaseolus lunatus* L.) as revealed by chloroplast DNA (cpDNA) variations. *Genet. Resour. Crop Evol.* 48: 437–445.

Freytag, G. F., and Debouck, D. G. (2002). Taxonomy, Distribution, and Ecology of the Genus *Phaseolus* (Leguminosae-papilionoideae) in North America, Mexico and Central America. Botanical Research Institute of Texas (BRIT), Forth Worth, TX, USA. pp. 298.

Fukunaga, K., Hill, J., Vigouroux, Y., Matsuoka, Y., Sánchez, G. J., Liu, K., Buckler, E. S., and Doebley, F. (2005). Genetic diversity and population structure of teocintle. *Genetics*. 169: 2241-2254.

Gaut, S. B. (2014). The complex domestication history of the common bean. *Nature Genetics*. 46: 663-664.

Gepts, P. (1990). Biochemical evidence bearing on the domestication of *Phaseolus* (Fabaceae) beans. *Economic Botany*, 44: 28-38.

Gepts, P. (2008). “Tropical environments, biodiversity and the origin of crops”. In: Moore P., Ming R. (eds). Genomics of tropical crop plants, Springer, New York.

Gutiérrez-Salgado, A., Gepts P., and Debouck D. G. (1995). Evidence for two gene pools of the Lima bean, *Phaseolus lunatus* L., in the Americas. *Genetic Resources and Crop Evolution*, 42: 15-28.

Hancock, F. J. (2012). Plant Evolution and the Origin of Crop species. *Cab Intl. Third edition*.

Harlan, J. R. (1975). Geographic patterns of variation in some cultivated plants. *Journal of heredity,* 66:184-191.

Harlan, J. R. (1992). Crops and Man. American Society of Agronomy, Crop Science Society of America, Madison, Wisconsin, USA.

Harlan, J. R. (1995). The living fields. Cambridge Univ Press, Cambridge.

Hawkes, J. G. (1983). The diversity of crop plants. Cambridge, MA: Harvard University Press.

Kaplan, L. (1965). Archaeology and domestication in American Phaseolus (beans). Economic Botany, 19:358-368.

Kaplan, L., and Lynch, T. F. (1999). *Phaseolus* (Fabaceae) in archaeology: AMS radiocarbon dates and their significance for pre-colombian agriculture. *Economic Botany*, 53:261-272.

Kirchhoff, P. (1943). Mesoamerica. Acta Americana. *Revista de la Sociedad de Antropología y Geografía*, 1:92–107.

Kwak, M., and Gepts, P. (2009). Structure of genetic diversity in the two major gene pools of common bean (*Phaseolus vulgaris* L., Fabaceae). *Theoretical and Applied Genetics*, 118: 979-992.

Kwak, M., Kami, J., and Gepts, P. (2009). The putative mesoamerican center of domestication of *Phaseolus vulgaris* L. is located in the Rio Lerma-Santiago basin of Mexico. *Crop Science*, 49: 554-563.

Lioi, L., and Galasso, I. (2002). Oligonucleotide DNA fingerprinting revealing polymorphism in *Phaseolus lunatus* L. *Genetic Resources and Crop Evolution,* 49: 53–8.

Lioi, L., Lotti, C., and Galasso, I. (1998). Isozyme diversity, RFLP of the rDNA and phylogenetic affinities among cultivated Lima beans, *Phaseolus lunatus* (Fabaceae). *Plant Systematics and Evolution*, 213: 153–64.

Lopéz-Soto, J. L., Ruiz, C. J. A., Sánchez, G. J. J., and Lépiz, I. R. (2005). Climatic adaptation of 25 wild bean species (*Phaseolus* spp.) in Mexico. *Revista Fitotecnia Mexicana*, 28:221-230.

Machuca-Chavéz, P., Colunga-García Marín, P., and Zizumbo-Villarreal, D. (2010). Introducción y difusión temprana de recursos fitogenéticos en la región Balsas-Jalisco durante el siglo XVI: una perspectiva agrohistórica. *Revista de Geografía Agrícola.* 44: 77-96.

Mackie, W. W. (1943). Origin, dispersal, and variability of the Lima bean, *Phaseolus lunatus. Hilgardia,* 15: 1-29.

MacNeish, R. S. (1967). Mesoamerican archaeology. *Biennial Review of Anthropology,* 5:306–331

MacNeish, R. S. (1992). The Origins of Agriculture and Settled Life. University of Oklahoma Press, Norman.

MacNeish, R. S., and Eubanks, E. M. (2000). Comparative analysis of the Rio Balsas and Tehuacan models for the origin of maize. *Latin American Antiquity,* 11:3–20.

Maquet, A., Zoro Bi, I., Delvaux, M., Wathelet, B., and Baudoin, J. P. (1997). Genetic structure of a lima bean base collection using allozyme markers. *Theoretical and Applied Genetics*, 95:980-991.

Martínez-Castillo, J., Zizumbo-Villarreal, D., Perales-Rivera, H., and Colunga-GarcíaMarín, P. (2004). Intraspecific diversity and morphophenological variation in *Phaseolus lunatus* L. from the Yucatan Peninsula, Mexico. *Econ Bot,* 58(3): 354 380.

Martínez-Castillo, J., Zizumbo-Villarreal, D., Gepts, P., Delgado-Valerio, and P., Colunga-GarcíaMarín, P. (2006). Structure and genetic diversity of wild populations of Lima bean (*Phaseolus lunatus* L.) from the Yucatan Peninsula, Mexico. *Crop Science*, 46: 1071–1080.

Martínez-Castillo, J., Zizumbo-Villarreal, D., Gepts, P., and Colunga-GarcíaMarín, P. (2007). Gene flow and genetic structure in the wild–weedy–domesticated complex of *Phaseolus lunatus* L. in its Mesoamerican center of domestication and diversity. *Crop Science*, 47: 58-66.

Martínez-Castillo, J., Colunga-GarcíaMarín, P., and Zizumbo-Villarreal, D. (2008). Genetic erosion and in situ conservation of Lima bean (Phaseolus lunatus L.) landraces in its Mesoamerican diversity center. *Gen Resour Crop Evol*, 55: 1065-1077.

Martínez-Castillo, J., Camacho-Pérez L., Villanueva-Viramontes S., Andueza-Noh R. H., and Chacón Sánchez, M. I. (2014). Genetic structure within the Mesoamerican gene pool of wild *Phaseolus lunatus* L. (Fabaceae) from Mexico as revealed by microsatellite markers: Implications for conservation and domestication of the species. *American Journal of Botany,* 101(5):851-864.

Matos-Moctezuma, E. (1994). Mesoamérica. In: Manzanilla, L., López-Luján, L., eds. Historia antigua de México. Vol. I. El México antiguo, sus áreas culturales, los orígenes y el horizonte Preclásico. México: Consejo Nacional para la Cultura y las Artes, Instituto Nacional de Antropología e Historia, Universidad Nacional Autónoma de México, Porrúa.

Matsuoka, Y., Vigouroux, Y., Goodman, M.M., Sanchez, G.J., Buckler, E., and Doebley, J. (2002). A single domestication for maize shown by multilocus microsatellite genotyping. *Proceedings of the National Academic of Sciences USA*, 99: 6080-6084.

Montes, C., Cardona, A., MacFadden, R., Moron, S. E., Silva, C. A., Restrepo-Moreno, S., Ramirez, D. A., Wilson, J., Farris, D., Bayona, G. A., Jaramillo, C., Valencia, V., and Flores, J. A. (2012). Evidence for middle Eocene and younger emergence in Central Panama: implications for Isthmus closure. *Geological Society of America Bulletin,* First published online January 13, 2012, doi: 10.1130/B30528.1

Motta-Aldana, J. R., Serrano-Serrano M. L., Hernández-Torres J., Castillo-Villamizar G., Debouck D. G., and Chacón S. M. I. (2010). Multiple origins of Lima bean landraces in the Americas: evidence from chloroplast and DNA polymorphisms. *Crop Sci.*, 50: 1773-1787.

Muñoz, C. L., Duque, C. M., Debouck, D. G., and Blair, W. M. (2006). Taxonomy of tepary bean and wild relatives as determined by amplified fragment length polymorphism (AFLP) markers. *Crop Science*, 46: 1744-1754.

Nanni, L., Bitocchi, E., Bellucci, E., Rossi, M., Rau, D., Attene, G., Gepts, P., and Papa, R. (2011). Nucleotide diversity of a genomic sequence similar to shatterproof (PvSHP1) in domesticated and wild common bean (*Phaseolus vulgaris* L.). *Theoretical and Applied Genetics*, 123: 1341-1357.

Nienhuis, J., Tivang, J., Skroch, P., dos Santos J. B. (1995). Genetic relationships among cultivars and landraces of lima bean *(Phaseolus lunatus* L.) as measured by RAPD markers. Journal of American Society *Horticultural Science*, 120:300- 306.

Piperno, D. R. and Pearsall, D. M. (1998). The origins of agriculture in the lowland Neotropics. San Diego, CA: Academic Press.

Piperno, D. R., Moreno, J. E., Holst, I., Lachniet, M., Jones, J. G., Ranere, A. J., and Castanzo, R. (2007). Late Pleistocene and Holocene environmental history of the Iguala Valley, central Balsas watershed of Mexico. *Proceedings of the National Academy of Sciences of the USA*, 104:11874–11881.

Pickersgill, B. (1989). Cytological and genetical evidence on the domestication and diffusion of crops within the Americas. In: Foraging and Farming: The Evolution of Plant Exploitation. Harris D.R., Hillman G.C., (eds). Unwin Hyman, London.

Rzedowski, J. (1993). Diversity and origins of the phanerogamic flora of Mexico. In: Ramamoorthy T., Bye, P., Lot R. A., FaJ, A. eds. Biological diversity of Mexico. New York, NY: Oxford University Press, 129–144.

Sanjur, O. I., Piperno, D. R., Andrés, T. C., and Wessel W. B. (2002). Phylogenetic relationships among domesticated and wild species of *Cucurbita* (Cucurbitaceae) inferred from a mitochondrial gene: Implications for crop plant evolution and areas of origin. *Proceedings of the National Academy of Sciences USA,* 99: 535-540.

Serrano-Serrano, M. L., Hernandez-Torres, J., Castillo-Villamizar, G., Debouck, D. G., and Chacón S. M. I. (2010). Gene pools in Wild Lima bean (*Phaseolus lunatus* L.) from the Americas: Evidences for an Andean origin and past migrations. *Mol. Phylogenet. Evol*, 54: 76-87.

Serrano-Serrano, M., Andueza-Noh, R., Martínez-Castillo, J., Debouck, D. G., and Chacon, M. I. (2012). Evolution and domestication of Lima bean (*Phaseolus lunatus* L.) in Mexico: evidence from ribosomal DNA. *Crop Science,* 52:1698-1712.

Schmutz, J., McClean, P. E., Mamidi, S., et al. (2014). A reference genome for common bean and genome-wide analysis of dual domestications. *Nature genetics,* 46(7), 707-713.

Smartt, J. and Simmonds, N. (1995). Evolution of crop plants. Wiley-Blackwell, second edition.

Smith, B. D. (2005). Reassessing Coxcatlan cave and the early history of domesticated plants in Mesoamerica. *Proceedings of the National Academy of Sciences USA*, 102:9438–9445

Sparvoli, F., Lanave, C., Santucci, A., Bollini, R., and Lioi, L. (2001). Lectin and lectin-related proteins in Lima bean (*Phaseolus lunatus* L.) seeds: biochemical and evolutionary studies. *Plant Mol. Biol.*, 45, 587–597.

Toledo, V. M., and Ordóñez, M. J. (1993). Thc biodiversity of Mexico: a review of terrestrial habitats. In: Ramamoorthy, T. P., Bye, R. A., Lot, A. eds. Biological diversity of Mexico. New York, NY: Oxford University Press. 757–777.

Toledo, V. M. (2000). Biodiversity and indigenous peoples. In: Levin SA, ed. Encyclopedia of biodiversity. Vol. 3. San Diego, CA: Academic Press. 451–463.

Vavilov, N. I. (1951). The origin, variation, immunity and breeding of cultivated plants. *Chronica Botanica*, 13: 1–16.

Villaseñor, J. L. (2003). Diversidad y distribución de las Magnoliophyta de México. *Interciencia,* 28: 160–167.

Zizumbo-Villarreal, D., and Colunga-GarciaMarín, P. (2010). Origin of agriculture and plant domestication in west Mesoamerica. *Genetic Resources and Crop Evolution*, 57: 813–825.

In: Phaseolus Lunatus
Editors: A. S. Ferreira de Araujo et al.
ISBN: 978-1-63482-494-1

Chapter 4

PHENOLOGY, VEGETATIVE DEVELOPMENT AND POLLEN VIABILITY

***Maria da Conceição Prado de Oliveira*[*1]*,**
***Ana Clara de Sousa Meirelles*[2]*,**
***Ângela Celis de Almeida Lopes*[1]**
***and Marcones Ferreira Costa*[3]**

[1] Departamento de Biologia, Universidade Federal do Piauí, Teresina, Brazil
[2] Bolsista de Desenvolvimento Cientifico Regional do CNPq - Nível C, Universidade Estadual de Maringá, Maringá, Brazil
[3] Programa de Pós Graduação em Genética e Melhoramento, Universidade Federal do Piauí, Teresina, Brazil

ABSTRACT

The species *Phaseolus lunatus* L., widely known as lima bean originated in Latin America and has great potential for agricultural development. Germination usually occurs between the sixth and tenth day after sowing the seeds, the cotyledons emerge between the ninth and the fourteenth day, and the first leaves emerge unifoliate between the

* Corresponding Author address Dr. Maria Conceição Prado Oliveira, Federal University of Piauí, Teresina, PI (Brazil); Tel.: +55-86-3215-5700. E-mail address: pradoliveira@hotmail.com

thirteenth and the fourteenth day. Regarding the pattern of flowering, this species presents daily production of many flowers per plant with high synchrony in a single season. The first flower buds appear between 36 and 52 days and the first flower between 60and46 days. The fruits appear between 51 and 74 days and the full maturity of pods runs between 84 and 97 days. The lima bean pollen comes in a circular shape with three pores opening and smooth surface, while maintaining a high rate of pollen viability.

Keywords: Environmental factors, pollen, reproductive development

INTRODUCTION

Phenology is the study of the influence of seasonal climate variations and edaphic and ecological conditions on the life cycle of a plant, especially on the reproductive and vegetative organs [14]. Knowledge and understanding of the phenological patterns of plants is essential to promote genetic improvements and manage and domesticate species. In addition, such information can explain the relationships between plants and their environment and can contribute to the understanding of pollination-related mechanisms.

Certain plant species display extreme variations at the beginning and end of their reproductive period, and the species can begin flowering approximately 30 days after germination and flower for 18 to 30 days, whereas others do not flower for 90 days or more after germination [15] and only flower for a few days.

Factors such as water stress, excess phosphate fertilisation, high temperatures, low light, inadequate photoperiods, and biotic factors (such as pests, disease and genetic factors) contribute to variations in the duration of the reproductive period and low ratios of fruits/flowers [6]. Biological and environmental factors can also have a direct effect on plant vegetative growth. Studies of seed morphology and germination and the early stages of a plant's life cycle are therefore important to the understanding of plant ecology.

In Brazil, several *native* plants, including *Phaseolus lunatus* L., have the potential to *contribute* to the agricultural development of the country [13].*P. lunatus* L. is commonly known as the lima bean or butter bean (feijão-fava, feijão- de-lima, fava lima, or fava, in Brazil) [11], and it belongs to the family Fabaceae, which is one of the largest plant families. This family has a wide geographical distribution and includes other food crop species, such as soya bean (*Glycine max*), pea (*Pisum sativum*), alfalfa (*Medicago sativa*) and common bean/*P. vulgaris*.

Lima beans are native to Latin America but are widely studied in temperate or sub-tropical regions of other continents and display excellent adaptability to humid and sub-humid areas of tropical zones. In Brazil, lima beans are grown throughout most of the country but are economically important in only certain states, such as Paraíba, Piauí and Maranhão. Lima beans are difficult to obtain because their production is mostly casual, with the plant intercropped with other crops in small scattered areas.

According to the literature, the reproductive biology of *Phaseolus lunatus* L. is already known. However, data on its floral phenology are practically non-existent. Information on pollen viability is fundamental in studies of the reproductive biology of plants and can be used in practical applications in genetic conservation, such as pollen monitoring, to guarantee fertilisation.Such data are important parameters in gene flow analyses.

Studies of pollen viability and morphology are of great importance for genetic improvement programs because this information can be used to obtain potentially promising selections through artificial crossings [7]. Pollen viability directly influences the success of fertilisation because the size of the fruit, number of seeds, and percentage of fruit increase if a larger quantity of viable pollen than required is deposited on the stigma

Pollen viability can be determined using direct methods, such as the induction of *in vitro* [10] and *in vivo* germination, or indirect methods based on cytological parameters, such as pollen staining [1]. These studies can be used to overcome genetic and morphological barriers that may occur during the hybridisation process.

PHENOLOGY AND VEGETATIVE DEVELOPMENT OF *P. LUNATUS* L.

The authors of the current chapter studied the vegetative and reproductive development of different *P. lunatus* L. genotypes under greenhouse and laboratory conditions. The experiments were performed in two stages.

The first stage of the study was conducted in a greenhouse located at the experimental area of the Department of Plant Science of the Federal University of Piauí (Universidade Federal do Piauí - UFPI). For this stage, 96 genotypes of *P. lunatus* L. from the UFPI Lima Bean Germplasm Bank (UFPI

465, UFPI 463 and UFPI 482) were used. The sowing was performed by placing four seeds per bag in a completely randomised experimental design, with five replicates per genotype. The plant growth and phenology were monitored between March and September 2006 between 06:00 and 11:00.

Germination of the *P. lunatus* genotypes occurred between day 6 and 10 after sowing. The cotyledons emerged between day 9 and 14, the first unifoliate leaves emerged between day 13 and 14, the first trifoliate leaf emerged between day 16 and 18, the second trifoliate leaf emerged between day 21 and 24, and the third trifoliate leaf emerged between day 23 and 26.

The primordia of the secondary branch emerged between day 27 and 28, the first leaf of the secondary branch emerged between day 29 and 30, the second leaf emerged between day 33 and 35, and the third leaf emerged between day 35 and 40 (Figures 1 and 2).

Flowering of *P. lunatus* occurred between April and July 2006, which corresponded to the rainy season, and the peak flowering occurred in the beginning of May, which suggests that this species is best adapted to mild temperatures.

This species displays a flowering pattern that includes the daily production of numerous flowers per plant, and there is high flowering synchronicity in a single season of the year. The availability of many flowers per inflorescence per day is common in the family Fabaceae and is advantageous for the species because it increases attractiveness to pollinators by increasing nectar availability. Approximately 90% of the flower buds of the studied genotypes are small, varying between 0.6 and 0.7 cm. A colour pattern was observed on the keel and banner of flowers; these structures were not pigmented in white flowers and were pigmented in light pink flowers. White flowers predominated (77%) (Figure 3).

During the reproductive stage, the first flower buds emerged between day 36 and 52, and the first flower emerged between day 46 and 60. Fruit maturation began between day 51 and 74, 50% pod maturity was observed between day 67 and 85, and 90% pod maturity was observed between day 84 and 97 (Figure 4).

Anthesis was observed to be diurnal, and it occurred between 06:30 and 10:30. The tested genotypes displayed long flowering periods and late cycles (flowering to pod maturity). Because of the proximity to the UFPI beekeeping area, the flowers were visited by bees during flowering, with *Apis mellifera L.* being the most frequent visitor. The bees would fly over the inflorescences and land on one of the open flowers. The visits occurred over the entire monitoring period and peaked between 08:30 and 10:00.

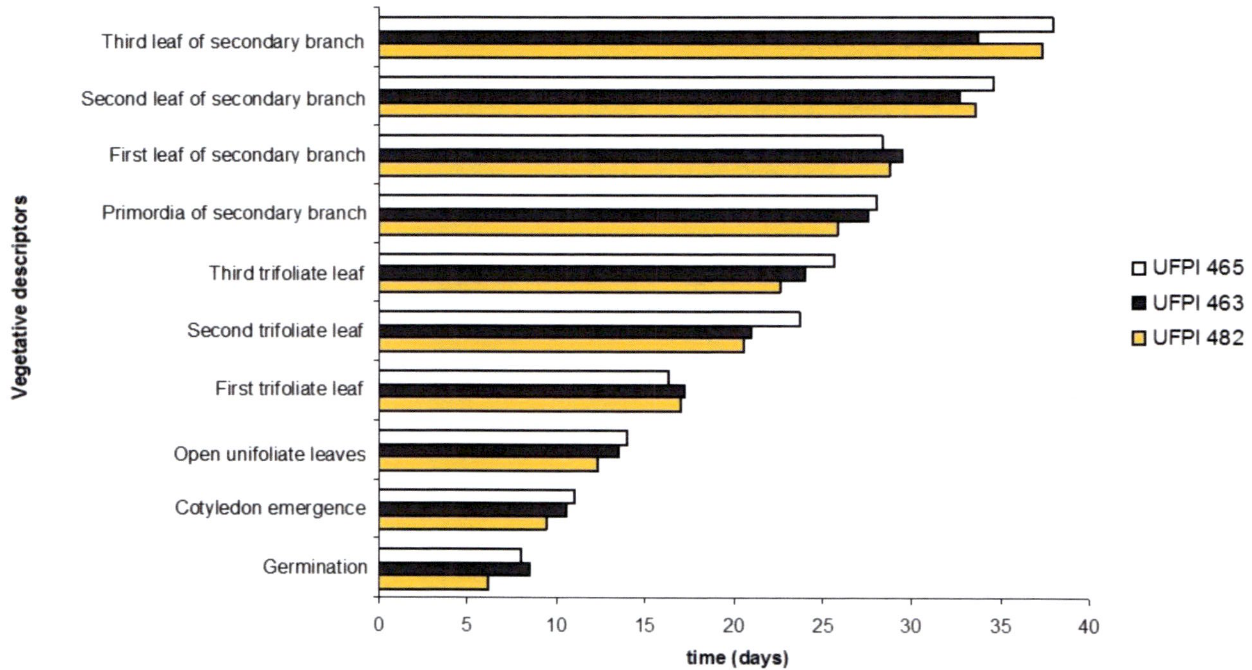

Figure 1. Stages of vegetative development of lima beans (*P. lunatus* L.).

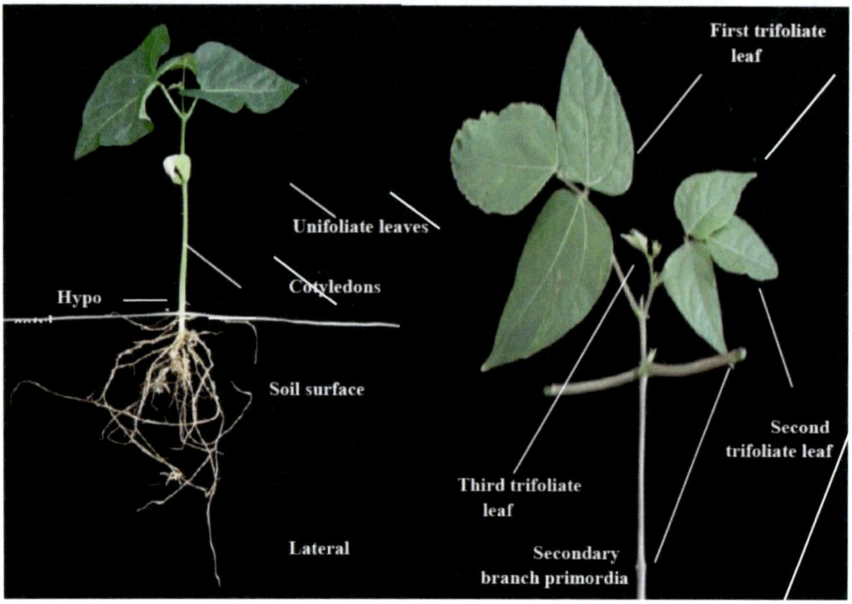

Figure 2. *Phaseolus lunatus* L. (A) seedling showing different vegetative stages (B).

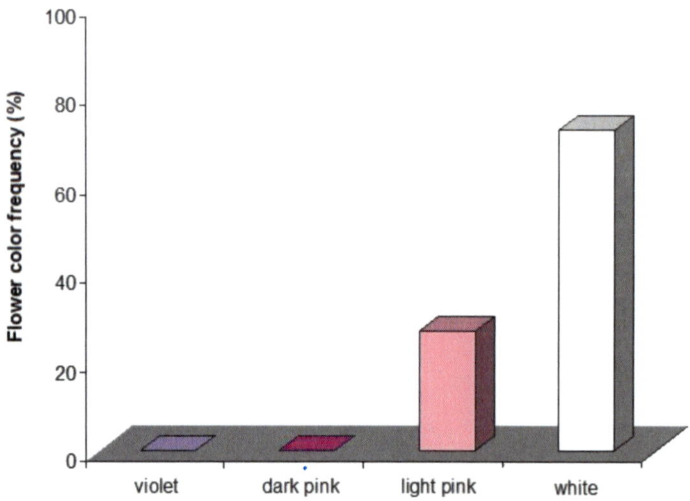

Figure 3. Colour of lima bean (*Phaseolus lunatus* L.) flowers, Teresina, Piauí (PI), 2006.

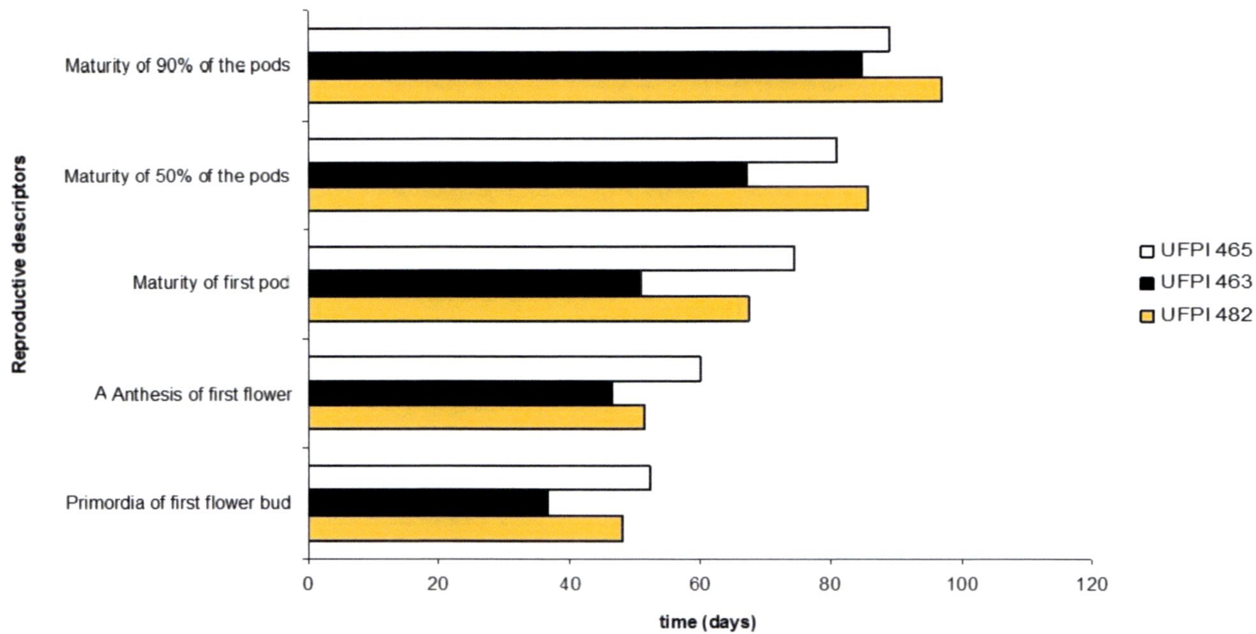

Figure 4. Stages of reproductive development of the lima bean (Phaseolus lunatus L.).

Lima beans have composite and axillary inflorescences, and the flowers are arranged in bunches with a variable number of buds per inflorescence. The flowers are hermaphroditic and papilionaceous with a gamosepalous calyx, and the corolla may have white, light pink, dark pink or violet petals. The androecium is formed by ten diadelphous stamens, anthers are bithecal and dorsifixed with longitudinal dehiscence, and there is a unilocular ovary presenting two to three ovules. A high number (~90%) of viable pollen grains was observed, and the grains were round with three pores and a smooth surface (Figure 5).

Figure 5. Lima bean (*Phaseolus lunatus* L.). Floral bud (A), pre-anthesis flower bud (B), flower in anthesis (C), flower at the end of anthesis (D), inflorescence cluster with several floral buds (E) and viable pollen grain (F) displaying three pores and a smooth surface. Teresina, PI (2006).

The second stage of the study was conducted at the Laboratory of Plant Ecophysiology of the UFPI Biology Department. At this stage, the development of three *P. lunatus* genotypes (UFPI 465, UFPI 463 and UFPI 482) was investigated under laboratory conditions.

The germination morphology and initial growth of the three tested genotypes were studied for 10 days. A total of 100 seeds per genotype were placed into 30 boxes lined with vermiculite, randomly distributed into different lots, and maintained on the laboratory bench at 25±2°C under continuous light. On days 1, 2, 4, 6, 8 and 10, a box of each genotype was removed for morphological analysis of the seeds and seedlings and quantification of the root and shoot growth. Following the analysis, the plant material was photographed using a digital camera.

The seed length and width were measured before and after imbibition using a pachymeter with 0.1 mm precision.

The length was measured from the base to the apex, and the width was measured along the midline of the seed. In the laboratory, the germination (radicle emergence) of *P. lunatus* began 48 hours after sowing for the three genotypes. In the germination tests, the genotypes UFPI 465 and UFPI 482 displayed a 50% increase of seed size 48 hours after sowing, although the same pattern was not observed for width. The genotype UFPI 463 varied little in size compared with the other genotypes tested. These results suggest that seeds from genotype UFPI 463 imbibe less water before germination than the other genotypes tested (Figures 6, 7 and 8).

Four days after sowing, seedlings from the three genotypes that were already presenting hypocotyls emerged from the substrate and cotyledons (enclosed in the testa) exposed above the soil surface. This result confirms that *P. lunatus* germination is epigeous, which was previously reported by Silva and Costa (2003). Within five to six days, the cotyledons became free from the testa (Figure 9), although they did not open completely as observed in the field.

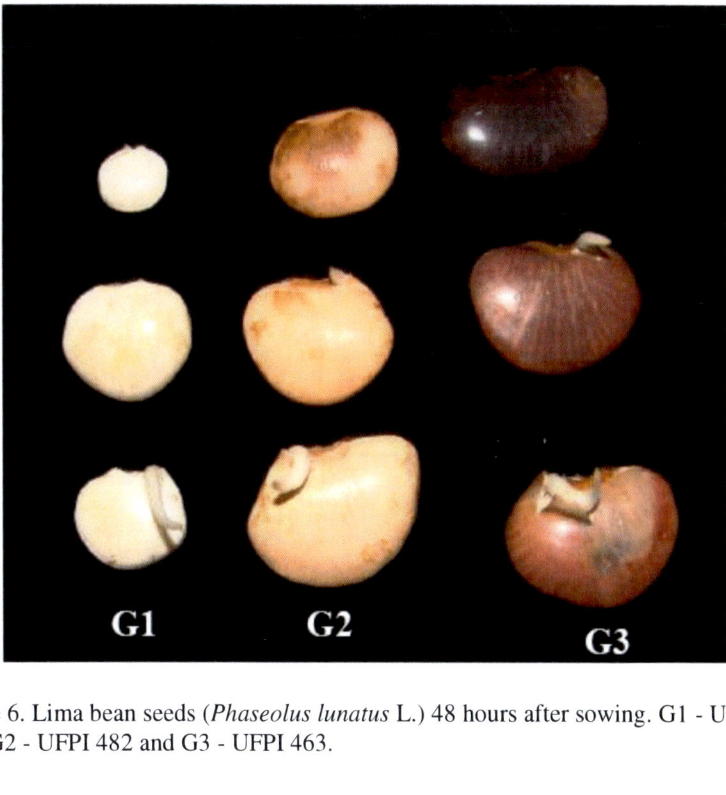

Figure 6. Lima bean seeds (*Phaseolus lunatus* L.) 48 hours after sowing. G1 - UFPI 465, G2 - UFPI 482 and G3 - UFPI 463.

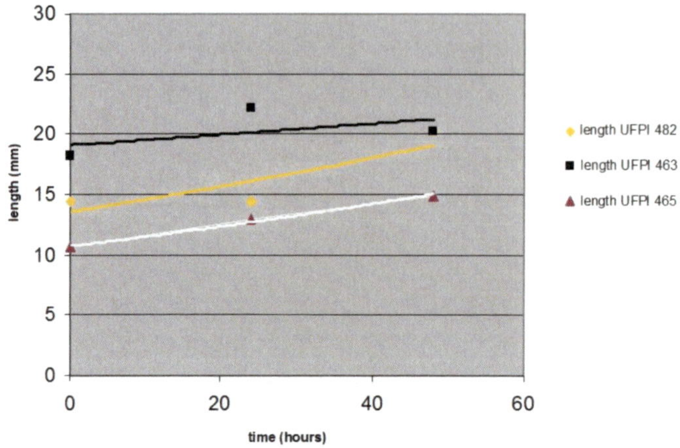

Figure 7. Length of the lima bean (*Phaseolus lunatus*) genotypes during imbibition.

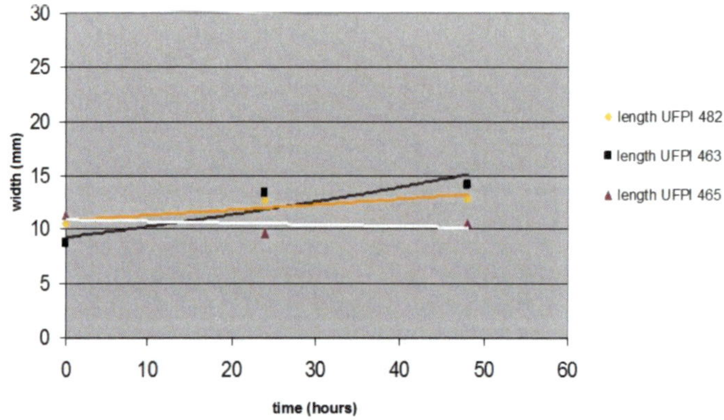

Figure 8. Width of the lima bean *(Phaseolus lunatus)* genotypes during imbibition.

Figure 9. (Continued).

Figure 9. Shoot and root of the lima bean (*Phaseolus lunatus*) genotypes G1 - UFPI 482 and G2 - UFPI 463.

The shoot growth and root growth of *P. lunatus* seedlings are shown in Figures 10 and 11. Genotype UFPI 463 exhibited higher shoot growth than UFPI 482 and UFPI 465; no significant differences were observed for root growth.

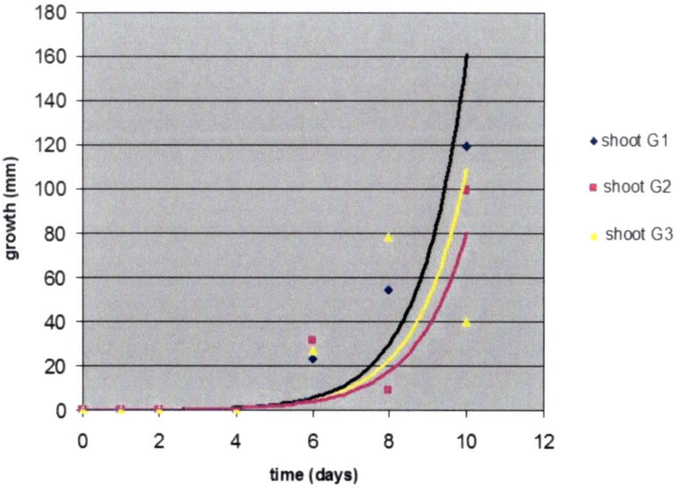

Figure 10. Shoot growth of the lima bean (*Phaseolus lunatus*) genotypes evaluated under laboratory conditions.

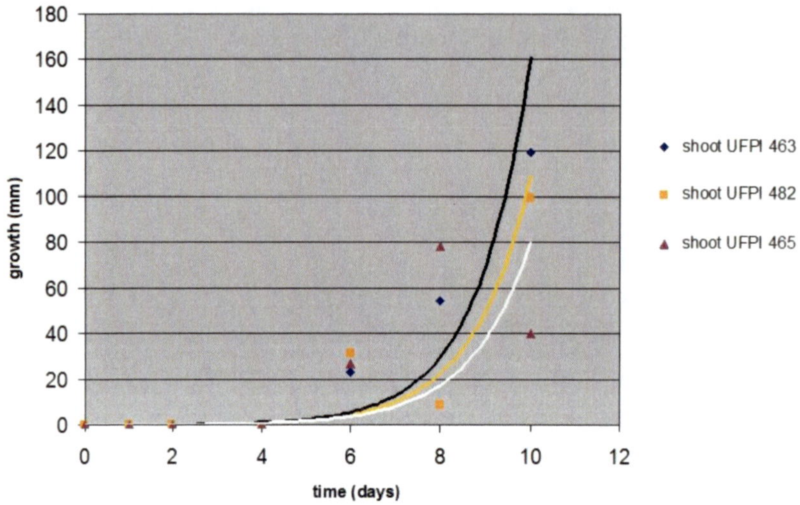

Figure 11. Root growth of the lima bean (*Phaseolus lunatus* L.) genotypes evaluated under laboratory conditions.

The results obtained in the laboratory suggest that the UFPI 463 genotype imbibes less water than the other genotypes tested. No large variations were observed in root and shoot development among the genotypes.

POLLEN VIABILITY OF *P. LUNATUS* L.

The pollen viability of *P. lunatus* L. was tested to evaluate the degree of viability and fecundity of the species. Five lima bean genotypes were obtained from the UFPI Active Germplasm Bank (UFPI-465, UFPI-728, G25165, G26200, UFPI-666). The genotypes had contrasting traits, especially for growth habit, size and precociousness. The genotypes originated from Piauí, Maranhão, Paraíba and CIAT (*International Centre for Tropical Agriculture*) in Colombia. Inflorescences in anthesis were collected at the beginning of flowering, placed in 70% ethanol and stored at -10ºC until the slides were prepared. No differences in the time of floral bud opening were observed among the different genotypes. Floral bud opening occurred over an entire day beginning at approximately 09:00 and continuing until 18:30, even at times when the temperature was slightly higher. A higher incidence of flower opening was observed between 09:30 and 11:00 and between 16:30 and 18:30 (Figures 12 and 13).

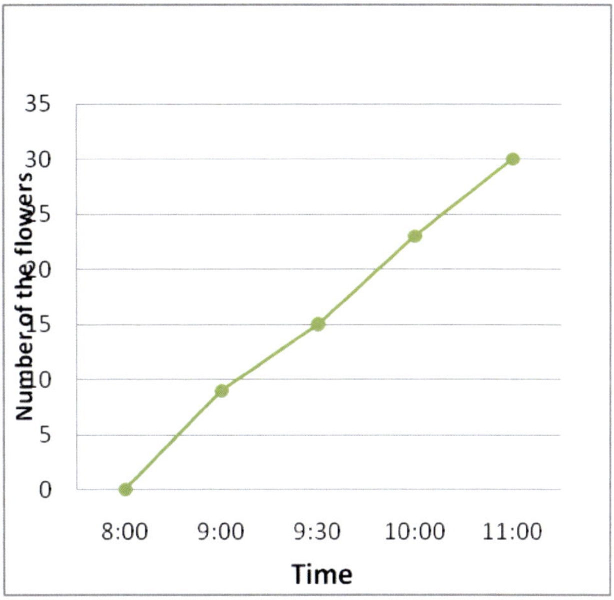

Figure 12. Time of opening of lima bean flowers during the morning.

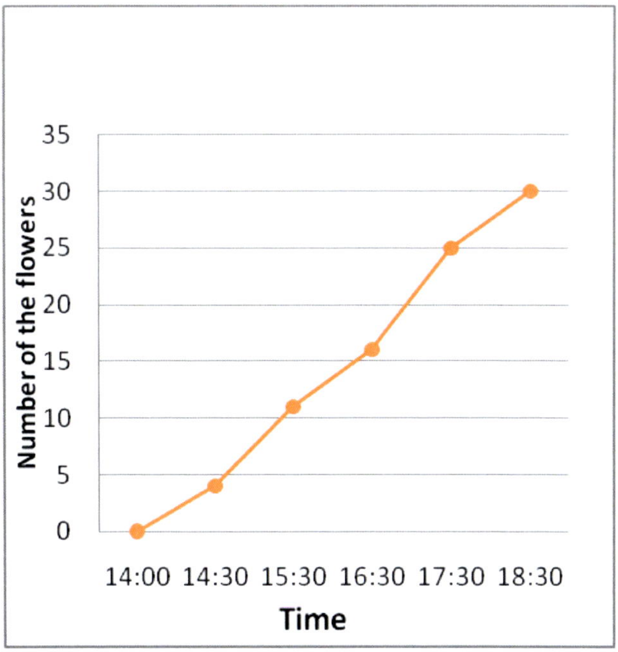

Figure 13. Time of opening of lima bean flowers during the afternoon.

The pollen viability was tested using an acetocarmine stain [4], which indicates pollen viability when red.

The slides were prepared according to the following method: anthers were cross-sectioned and macerated using a scalpel to release the pollen grains onto 10 μL of acetocarmine. The material was covered with a cover slip and observed under an optical microscope.

A random sample of stained pollen grains was obtained using the sweeping method. Ten slides were prepared per lima bean genotype, and 300 pollen grains were counted per slide, for a total of 3000 pollen grains per genotype.

The data were subjected to an analysis of variance and then Tukey's test at $p<0.05$ using the software GENES [2]. The pollen grains that were stained red were considered viable, and unstained pollen were considered unviable (Figure 14).

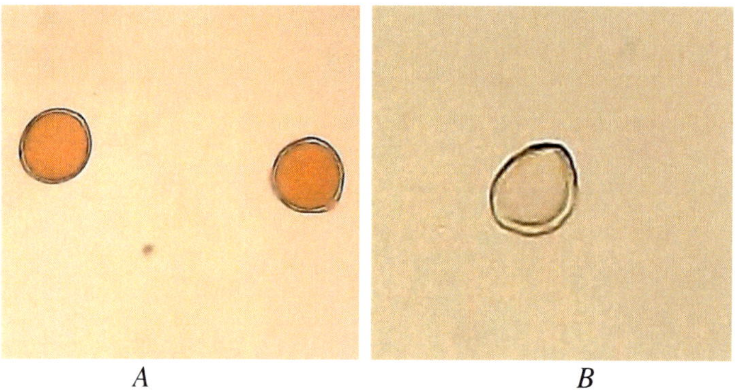

Figure 14. Pollen grains of lima beans. A- stained pollen grains considered viable; B- unstained pollen grain considered unviable.

The prevalence of sterile/unviable pollen grains was not observed for any of the genotypes studied. In contrast, high percentages of viable pollen grains were observed.

The pollen viability of the genotypes that were tested and evaluated using acetocarmine staining was higher than 88% (Table 1). Values higher than 70% indicate high viability, values between 31 and 69% indicate medium viability, and values below 30% indicate low viability [9]. Genotype G25165 presented the highest mean pollen viability percentages, which were significantly different ($p<0.05$) from those of genotype UFPI-666, which exhibited the lowest mean.

Table 1. Percentage of pollen viability for the lima bean (*Phaseolus lunatus* L.) genotypes that were tested using acetocarmine for fixed pollen grains

Genotype	Mean
UFPI-465	90.05 ab
UFPI-728	91.76 ab
G25165	94.07 a
G26200	92.64 ab
UFPI-666	89.3 b

*Means followed by different letters were significantly different according to Tukey's test at $p<0.05$.

Because meiosis, stability and production of viable gametes are directly associated with reproduction and maintenance of this species, the high viability of the pollen indicates that the tested lima bean genotypes exhibit normal meiosis that will result in viable pollen grains capable of germinating in the stigma and of fertilising and generating viable pods.

FINAL CONSIDERATIONS

Environmental conditions are known to vary between locations, which affects plant responses during their life cycle. Studies of the vegetative and reproductive development of plants under different conditions (natural and controlled in the greenhouse and laboratory) are therefore fundamental to understanding the dispersal patterns of certain species, especially food crop species such as *P. lunatus* L. (lima bean), which is common in northern Brazilian cuisine.

P. lunatus L. genotypes from the UFPI Germplasm Bank (UFPI 465, UFPI 463, UFPI 482) exhibit fast germination under laboratory conditions. The seeds from genotype UFPI 463 imbibed less water for germination than the seeds of the other tested genotypes. The root growth and shoot growth were rapid, and no large differences were observed among genotypes.

Under greenhouse conditions, the life cycle of the tested genotypes lasted approximately 120 days. Between days 84 and 97, almost all the pods could already be harvested, which may be considered an excellent result for a species that produces grain for human consumption.

The lima bean displays high pollen viability, indicating regular meiosis and a normal meiotic index, which are important factors in genetic improvement programs, especially when artificial crossings are used because high viability reflects the potential and fecundity of male gametes and suggests subsequent fertilisation.

References

[1] Abdelgadir, H.A.; Johnson, S.D. & Van Staden, J. Pollen viability, pollen germination and pollen tube growth in the biofuel seed crop *Jatropha curcas* (Euphorbiaceae). *South African Journal of Botany*, p.79: 132-139, 2012.

[2] Cruz, C. D. *Programa genes: análise multivariada and simulação* [Genes Program: multivariate analysis and simulation]. Viçosa, MG: Ed. UFV, 175p, 2013.

[3] Fakhim, R.S.H.; Hajilou, J. & Zaare, N.F. Pollen germination and pistil performance in several Iranian peach cultivars. *International Journal AgriScience*, p. 1: 71-77, 2011.

[4] Kearns, C.A. & Inouye, D. *Techniques for pollinations biologists.* Niwot, Colorado, University Press of Colorado, 1993.

[5] Silva, H. T. da. Caracterização botânica das espécies silvestres *Phaseolus L.* (Leguminosae) [Botanical characterisation of wild species of *Phaseolus L.* (Leguminosae)]. In: SILVA, H. T. DA.; COSTA-SANTO, A. O.; GOIÁS, A. DE (Ed). Brasília. *Embrapa Arroz e Feijão*, p.40, 2003.

[6] Summerfield, R. J.; Pate, J. S. Robert, E. H. The Physiology Of Cowpea. In: Singh, S. R.; Rachie, K. O. (Ed). Cowpea research, production and utilization. Great Britain: A. *Wiley-Interscience Publication*, p. 65-102, 1985.

[7] Soares, T. L; Jesus, O. N De; Souza, E. H. De; Santos-Serejo, J.A Dos; Oliveira, E. J DE. Morphology and viability of pollen grains from passion fruit species (*Passiflora* spp.). *Acta Botanica Brasílica*, v. 27, p. 779-787, 2013.

[8] Rigigamoto, R.R; Tyagi, A.P. Pollen fertility status in coastal plant species of *Rotuma* Island. *S. Pac. J. Nat. Sci*, p. 20:30-33, 2002.

[9] Souza M. M.; Pereira, T. N. S.; Martins, E. R. Microsporogênese e microgametogênese associadas ao tamanho do botão floral e da antera e viabilidade polínica em maracujazeiro-amarelo (*Passiflora edulis* Sims f.

flavicarpa degener) [Microsporogenesis and microgametogenesis in relation to flower bud and anther size and pollen viability in yellow passion fruit (*Passiflora edulis* Sims f. flavicarpa degener)]. *Ciência Agrotécnica.*, Lavras. V.26, n.6, p.1209-1217, 2002.

[10] Sorkheh, K.; Shiran, B.; Rouhi, V.; Khodambashi, M.; Wolukau, J.N. & Ercisli, S. 2011. Response of in vitro pollen germination and pollen tube growth of almond (*Prunus dulcis* Mill.) to temperature, polyamines and polyamine synthesis inhibitor. *Biochemical Systematics and Ecology*, p. 39: 749-757, 2011.

[11] Vieira, C. Leguminosas de grãos: importância econômica na agricultura e na alimentação humana [Grain legumes: economic importance for agriculture and human feeding]. *Informe Agropecuário*, Belo Horizonte. v.16, n.174, p.5-11, 1992.

[12] Vieira, R. F. A cultura do feijão-fava [Cultivation of lima bean] -. *Informe Agropecuário*, Belo Horizonte, v. 16, n. 174, p. 30-37, 1992.

[13] Vilela-Morales, E. A.; Valois, A. C. C. Recursos genéticos vegetais autóctones e seus usos no desenvolvimento sustentável [Autochthonous plant genetic resources and their uses for sustainable development]. *Cadernos de Ciência & Tecnologia*, Brasília, v.17, n.2, p.11-42. 2000.

[14] Wielgolaski, F. E. Phenology in agriculture. *In*: H. Lieth (ed.). *Phenology and seasonality modeling*. Chapman & Hall, London, p. 369-381, 1974.

[15] Wien, H. C.; Summerfield, R. J. Cowpea (*Vigna Unguiculata* (L.) Walp.). In: Goldsworthy, P. R.; Fischer, N. M. (Ed.). *The physiology of tropical field crops*. Chichester: John Willey & Sons, p. 353-383, 1984.

In: Phaseolus Lunatus
Editors: A. S. Ferreira de Araujo et al.
ISBN: 978-1-63482-494-1

Chapter 5

CHEMICAL AND ORGANIC FERTILIZATION OF LIMA BEAN IN BRAZIL

Ademir Sérgio Ferreira de Araújo**, *Sandra Mara Barbosa Rocha*, *Dyego Leandro da Costa Monte* and *Nilza Silva Carvalho

Agricultural Science Center, Federal University of Piauí, Teresina, Piauí, Brazil

ABSTRACT

Lima bean (P. lunatus) is an important species of Phaseolus for human consumption in tropical regions and its seed is an important protein source for populations of South America, Africa and Mexico. The soil fertility and fertilization are important factors to improve lima bean yield. There is little information about soil fertilization for lima bean. Lima bean needs suitable quantity of nutrients such as N, P and K; thus, the chemical and organic fertilization should provide suitable soil condition to high level of plant yield.

The use of a liming and fertilization program can ensure success in the production of lima bean and provide income gains to farmers.

* Corresponding Author address Dr. Ademir Sergio Ferreira de Araújo. Agricultural Science Center, Federal University of Piauí, Teresina, PI (Brazil); Tel.: +55-86-3215-5740; fax: +55-86-3215-5743. Email address: asfaruaj@yahoo.com.br.

Education and the incentive to producers about the rational and economic use of lime and fertilizer is a practice that should be disseminated by agronomists and extension.

Keywords: Biological nitrogen fixation; legumes; rhizobia diversity

INTRODUCTION

Lima bean (*Phaseolus lunatus*) is an important species of plant for humans in tropical regions. This legume originated in Peru; and archeological evidence supports the hypothesis that lima beans were domesticated in Mesoamerica and South America (Salgado et al. 1995). Lima beans originated from two main groups: one from the lowlands of eastern South America, and another from the western Andes (Salgado et al. 1995). This species is predominantly autogamous, although it has outcrossing rates of up to 48% (Baudoin et al. 1998). It is characterized by an indeterminate, climbing growth habit, a prolonged flowering period, and production of a large number of pods (Zoro Bi et al. 2003).

The lima bean seed is considered a main crop and an important source of protein for people from South America, Africa, and Mexico. It also has great relevance in Brazil, mainly in the Northeast, where it is used as an alternative food and as an economic crop.

The research unit Embrapa Genetic Resources and Biotechnology has a large collection of germplasm from *P. lunatus*, with approximately 980 subsamples, although it has not been widely studied (Silva, 2011). *P. lunatus* seeds vary in size, shape, and color, and they range from flat-seeded types to the round-seeded potato types, which are cream, red, purple, brown, black, and various types of mottled colors (Vieira, 1992).

The seeds have high protein (210–260 $g \cdot kg^{-1}$) and carbohydrate (550–640 $g \cdot kg^{-1}$) contents, low fat (10–23 $g \cdot kg^{-1}$) levels and high fiber levels (32–68 $g \cdot kg^{-1}$), high levels of the minerals K, Zn, Ca, and Fe, and low levels of Na and P (Chel-Guerrero et al. 2012).

In Brazil, the lima bean is the second most economically important species of *Phaseolus* and shows a higher capacity of adaptation (Azevedo et al. 2003). According to IBGE (2013), Brazil produced approximately 14,951 tons of lima beans from an area of 37,521 ha.

The rustic quality of lima bean and its capacity to resist to long, dry periods are important characteristics for the semi-arid region of Northwest Brazil (Azevedo et al. 2003).

The soil fertility and fertilization are important factors to improve lima bean yield. There is little information about soil fertilization for lima bean. Lima bean needs suitable quantity of nutrients such as N, P and K; thus, the chemical and organic fertilization should provide suitable soil condition to high level of plant yield.

SOIL SAMPLING

Soil sampling is the first step involving in a soil analysis and fertilization. Although it is a simple step, the procedure should be done carefully aiming increasing the assurance of soil analysis.

In agriculture, a soil analysis determines nutrient and contaminated content, composition, and other characteristics such as the acidity or pH level. A soil analysis estimates soil fertility, or the expected growth potential of the soil which indicates nutrient deficiencies, potential toxicities from excessive fertility and inhibitions from the presence of non-essential trace minerals.

The test is used to mimic the function of roots to assimilate minerals. Soil analysis is often performed by commercial labs, targeting groups of compounds and minerals. Soil analysis often check for plant nutrients in three categories:

Major nutrients: nitrogen (N), phosphorus (P), and potassium (K)
Secondary nutrients: sulfur, calcium, magnesium
Minor nutrients: iron, manganese, copper, zinc, boron, molybdenum, chlorine

Fertilizer recommendations require an accurate sample of the field. The area should be representative of the field to be fertilized. Previous crop history, previous lime and fertilizer rates, and soil characteristics will cause variance in results if not accounted for while sampling.

Soil sampling should be done in similar field in soil type (slope, drainage, color or texture). About 20 individual cores of uniform size and depth collected from throughout the sampling zone or area should be composited in a clean, plastic bucket.

Usually, the soil samples is done from the surface to a depth of 20 cm. Mix these cores thoroughly and fill the sample bag. Label the bag clearly and use this same information and identification on the soil test information form.

SOIL ACIDITY

The properties of soils that relate to their acidity, indicated by pH measurement, are fundamental to understanding the nature and behavior of many soils. Processes of acidification are important in the management and productivity of agriculture and in environmental management.

The soil pH of a sample reflects the dissociation of weak and strong acids and bases that are associated with the solutions, salts and mineral and organic compounds that make up soil materials. Change in pH mostly results from nitrogen, carbon and sulfur cycle effects that result from natural biological processes in soils, and rates of change may be accelerated by human activities through agriculture and industry. Soil fertilization, also, can accelerate or alter acidification processes and in this case lime application should be recommend to improve the soil condition.

SOIL LIMING

Liming is the application of calcium- and magnesium-rich materials to soil.

This will neutralize soil acidity and increases activity of soil microorganisms and fertilizers efficiency. However, oversupply may result in harm to plant life.

The degree to which a given amount of lime per unit of soil volume will increase soil pH depends on the cation exchange capacity (CEC). Soils with low CEC will show a more marked pH increase than soils with high CEC. But the low-CEC soils will witness more rapid leaching of the added bases, and so will see a quicker return to original acidity unless additional liming is done. Over-liming is most likely to occur on soil which has low CEC, such as sand which is deficient in buffering agents such as organic matter and clay.

In Brazil, the main methods for soil liming recommended to Lima bean are:

1) Neutralization of exchangeable Al and increase of exchangeable Ca and Mg content. The quantity of calcareous should be calculated by:

   a) Rate (t/ha) = f x Al

or

   b) Rate (t/ha) = f x [2 – (Ca +Mg)]

Considering f = 1.5 (soil with clay content < 15 %); f = 2 (soil with clay content ranged from 15% to 35 %) e f= 2.5 (soil with clay content > 35%).

The final quantity of calcareous should be calculated considering:

Final quantity of calcareous (t/ha) = Rate (t/ha) x f

where: f =100/PRNT

2) Elevation of Base Saturation (BS). The farmers should elevate soil BS to 70% troughs the equation:

   c) Rate (t/ha) = (BS2 – BS1) x CECpH7/PRNT

where:

BS_2 = 70%;

BS_1 = actual soil BS;

CEC_{pH7} = Cation Exchange Capacity ($cmol_c/dm^3$);

PRNT = 100%;

The calcareous should be applied at least 2 months before and incorporated into the soil (20 cm).

Chemical Fertilization

Soil fertility refers to the ability of a soil to supply plant nutrients. The main plant nutrients used by lima bean is nitrogen, phosphorus and potassium.

Nitrogen is part of every living cell and usually increases plant growth more than any other element. Inside the plant, nitrogen is part of amino acids, which in turn make up proteins. Thus nitrogen is an important component of

DNA. It is also part of the chlorophyll molecule, thus it is important in photosynthesis.

Phosphorus serves as the currency of energy exchange within the plant itself. Thus it plays roles in photosynthesis, respiration, cell division, cell enlargement, and many other processes within the plant. It promotes early root formation, and improves the quality of many fruits and vegetables.

Potassium is somewhat of an enigma. It is essential for plant growth, however the exact functions within plants are not well understood. The primary function may be related to plant metabolism. It is known to be vital to photosynthesis, because it declines when potassium is deficient. Respiration increases when potassium is deficient.

Combined, increased respiration and decreased photosynthesis reduces the plant carbohydrate supply. Potassium is known to be essential for protein synthesis and enzyme activation.

Bioavailable nitrogen is the element in soil that is most often lacking for lima bean. The crop needs high quantity of phosphorus and potassium.

Organic Fertilization

Organic fertilizers are used in high quantities by the lima bean. Usually, the organic fertilizers derived from animal residues, such as cattle manure. This type of manure is rich in nitrogen, phosphorus and potassium, and provides numerous benefits to the lima bean. The cattle manure used into the soil improves its moisture-holding capacity. This allows you to water less frequently, as the roots of plants can use the additional water and nutrients whenever needed. Additionally, it will improve aeration, helping to break up compacted soils.

Composted cattle manure is another option for organic fertilization. This organic fertilizer contains beneficial bacteria, which convert nutrients into easily accessible forms so they can be slowly released without burning tender plant roots. Composting of cattle manure also produces about a third less greenhouse gases, making it environmentally friendly.

The recommendation of organic fertilizer for lima bean depends from the different type of fertilizer. Alves (2006) studied lima bean and found high plant yield after the organic fertilization with cattle manure (16 tons per ha) associated with chemical fertilizers.

SUGGESTIONS OF SOIL FERTLIIZATION FOR LIMA BEAN

Table 1. Soil fertilization with N, P_2O_5 and K_2O for Lima bean, without irrigation, according to soil analysis

Nutrient content	At planting	At cover
	----------------------- kg per ha -----------------------	
	Nitrogen (N)	
	20	-
	Phosphorus (P_2O_5)	
P (mg dm^{-3})		
< 11	60	-
11-20	40	-
> 20	20	-
	Potassium (K_2O)	
K ($cmol_c$ dm^{-3})		
< 0.12	45	-
0.12 – 0.23	30	-
> 0.23	15	-

Table 2. Soil fertilization with N, P_2O_5 and K_2O for Lima bean, with irrigation, according to soil analysis

Nutrient content	At planting	At cover
	----------------------- kg per ha -----------------------	
	Nitrogen (N)	
-	20	40
	Phosphorus (P_2O_5)	
P (mg dm^{-3})		
< 6	80	-
6 - 10	60	-
11 - 20	40	
> 20	20	-
	Potassium (K_2O)	
K ($cmol_c$ dm^{-3})		
< 0.08	80	-
0.08 – 0.15	60	-
0.16 – 0.25	40	-
> 0.25	20	-

Final Considerations

The use of a liming and fertilization program can ensure success in the production of the lima bean and provide income gains to farmers. Education and the incentive to producers about the rational and economic use of lime and fertilizer is a practice that should be disseminated by agronomists and extension.

Acknowledgments

The authors are grateful to 'Conselho Nacional de Desenvolvimento Científico e Tecnológico' (CNPq-Brazil) for financial support of research with Lima bean.

All authors are supported by Research fellowships from CNPq-Brazil (Research Productivity).

References

Azevedo JN, Franco JD, Araújo ROC (2003) Composição química de sete variedades de feijão-fava. Teresina, Embrapa Meio-Norte, 4 pp.

Baudoin JP, Degreef J, Hardy O, Janart F, Zoro Bi I (1998) Development ofan *in situ* conservation strategy for wild Lima bean (Phaseolus lunatus L.) populations in the central valley of Costa Rica. In: Owens SJ, Rudall PJ (Eds) Reproduction Biology, Royal Botanic Garden Press, England, pp 417-426.

Chel-Guerrero L, Domínguez-Magaña M, Martínez-Ayala A, Dávila-Ortiz G, Betancur-Ancona D (2012) Lima Bean (Phaseolus lunatus) Protein Hydrolysates with ACE-I Inhibitory Activity. *Food Nutrit Sci.*, 3:511-521.

Salgado AG, Gepts P, Debouck DG (1995) Evidence for two gene pools of the Lima bean, Phaseolus lunatus L., in the Americas. *Gen. Res. Crop Evol.,* 42:15-28.

Silva RNO (2011) Diversidade genética em feijão-fava (Phaseolus lunatus L.) por marcadores morfoagronômicos e moleculares. Dissertation of Master, UFPI, 2011, p.175.

Vieira RFA (1992) Cultura do feijão-fava. *Inf. Agropec.*, 16:30-37.

Zoro Bi I, Maquet A, Baudoin JP (2003) Population genetic structure of wild Phaseolus lunatus (Fabaceae), with special reference to population sizes. *Am. J. Bot.*, 90:897-904.

In: Phaseolus Lunatus
Editors: A. S. Ferreira de Araujo et al.
ISBN: 978-1-63482-494-1

Chapter 6

BIODIVERSITY OF NITROGEN-FIXING NODULE BACTERIA ASSOCIATED WITH LIMA BEAN (*PHASEOLUS LUNATUS L.*) IN ITS DOMESTICATION CENTERS

Ernesto Ormeño-Orrillo[1,*], Doris Zúñiga Dávila[2], and Esperanza Martínez-Romero[1]

[1]Center for Genomic Sciences, Universidad Nacional Autónoma de México, Cuernavaca, Morelos, Mexico

[2]Microbial Ecology and Biotechnology Laboratory, Department of Biology, Universidad Nacional Agraria La Molina, Lima, Peru

ABSTRACT

Like other leguminous plants, the Lima bean (*Phaseolus lunatus*) establishes a mutualistic symbiosis with nitrogen-fixing bacteria commonly known as rhizobia. These bacteria elicit the formation of nodule structures on the roots where they become intracellular and fix atmospheric nitrogen for the plant. The Lima bean shows a preference to form nodules with slow-growing bacteria of the genus *Bradyrhizobium* unlike its congeneric species, the common bean (*Phaseolus vulgaris*), which prefer fast-growing rhizobial strains. *P. lunatus* was independently domesticated in the Andes and Mesoamerica by pre Columbian societies.

* Email: eormeno.orrillo@gmail.com, eormeno@ccg.unam.mx

Knowledge of the genetic biodiversity of Lima bean symbionts has increased in recent years with data obtained in its domestication areas. In Peru, Lima beans host *Bradyrhizobium yuanmingense*, *Bradyrhizobium paxllaeri*, *Bradyrhizobium icense*, and a yet unnamed genospecies. At least seven *Bradyrhizobium* genospecies have been found in *P. lunatus* nodules in Mexico. This rhizobial richness indicates a relative promiscuity of Lima bean in contrast to previous statements that this plant was a restricted host for nodulation. Interestingly, Mexican and Peruvian bradyrhizobia are distinct, perhaps in relation to the different plant gene pools from which domestication occurred in those geographic areas. Besides *Bradyrhizobium*, other rhizobial genera such as *Rhizobium*, *Sinorhizobium* and *Mesorhizobium* can be found in Lima bean nodules in its domestication centers and in other areas. Although effective nodulation with these genera has been reported, the ecological relevance of those associations is presently unknown.

Keywords: Pallar, ibes, Mexico, Peru, biological nitrogen fixation, biofertilizer, inoculant

INTRODUCTION

The Lima bean (*P. lunatus* L.) is one of five cultivated species of the leguminous genus *Phaseolus*. The other cultivated species are common bean (*Phaseolus vulgaris* L.), runner bean (*Phaseolus coccineus* L.), tepary bean (*Phaseolus acutifolius* A. Gray) and year bean (*Phaseolus polyanthus* Greenman). The Lima bean was independently domesticated from distinct wild gene pools in the Andean and Mesoamerican regions by pre Columbian societies (Gutiérrez Salgado et al., 1995). This species is the second most economically important species of the genus after common bean. In general, *Phaseolus* beans have been important sources of low-cost, high-quality proteins for human populations in the Americas for millennia, and for other people in places like Africa after their introduction from America.

As with other leguminous plants, Lima beans are able to establish a mutualistic symbiosis with nitrogen-fixing bacteria known as rhizobia (Santos et al. 2008). These bacteria inhabit soils and upon perception of flavonoids released by the roots activate the expression of nodulation (*nod*) genes, which are involved in the biosynthesis and secretion of molecules known as Nodulation (Nod) factors. Nod factors are perceived by the plant where they act as signals to trigger the morphogenesis of specialized root structures known as nodules. Inside nodules rhizobia express *nif* genes required for nitrogen fixation. In exchange for fixed nitrogen, rhizobia receive carbon sources to sustain the nitrogen fixation process. Leguminous plants can fulfill most of their nitrogen needs through this symbiosis. Rhizobia can be grown in synthetic media and formulated in products known as biofertilizer inoculants that can be applied to seeds or soil in order to ensure the formation of the symbiosis. In agriculture, rhizobial inoculants represent cost effective and ecological friendly replacements for chemical nitrogen fertilizers (Zúñiga-Dávila, 2007).

Effective use of rhizobia requires knowledge on their biodiversity and symbiotic characteristics. Biodiversity of rhizobia has been, and still is, intensively studied around the globe. Several genera of alpha and beta Proteobacteria include rhizobial members. Most published work on rhizobia associated with *Phaseolus* beans centers on symbionts of *P. vulgaris* bean because of its economic importance and its worldwide distribution. In contrast, relatively few studies on Lima bean rhizobia are available. In the 1930s, it was shown that rhizobia associated with Lima bean belonged to the so called "cowpea miscellany group", a heterogeneous assemblage of rhizobia able to establish symbiosis with cowpea (*Vigna unguiculata* (L.) Walp.) and other tropical leguminous species (Allen and Allen, 1939). Slow-growing rhizobial members of this group were later ascribed to the *Bradyrhizobium* genus of the alpha Proteobacteria (Jordan, 1984). Thies et al. (1991) reported that Lima bean was a relatively restricted host for nodulation when compared to other hosts like cowpea (*Vigna unguiculata*) and siratro (*Macroptilium atropurpuereum*).

Until 2006, reports on Lima bean rhizobia focused mainly on their performance as inoculant strains. Furthermore, most of those rhizobia were isolated in regions where the Lima bean was not native and none of them were characterized by molecular methods so their assignment to *Bradyrhizobium* or other rhizobial genera was based only on growth rate. In this chapter we will summarize the studies on biodiversity of the Lima bean rhizobia in the domestication sites of *P. lunatus*.

BIODIVERSITY OF LIMA BEAN RHIZOBIA IN THE ANDEAN REGION

Archeological and molecular evidence suggest that Lima bean was domesticated in Western South America (Northern Peru and Ecuador) from the Andean gene pool of *P. lunatus* (Gutiérrez Salgado et al., 1995). Nowadays, this crop is cultivated in several countries in South America. In Peru, Lima bean is cultivated mainly in the central and south subtropical coastal regions.

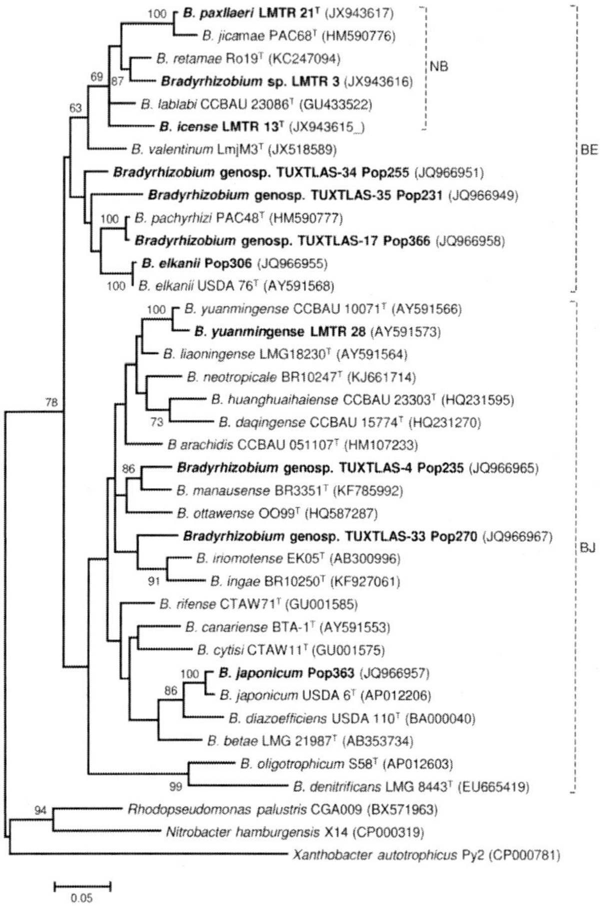

Figure 1. Phylogenetic tree showing the relationships between Lima bean bradyrhizobial symbionts (shown in bold) and all currently described *Bradyrhizobium* species based on *recA* gene sequences. Type strains are indicated with a superscript letter T. Phylogenetic branches indicated with dotted brackets: NB, branch first described by Ormeño-Orrillo *et al.* (2006); BE, *B. elkanii* 16S rRNA phylogenetic superclade; BJ, *B. japonicum* 16S rRNA phylogenetic superclade. The tree was constructed with the maximum likelihood method using GTR+I+G as evolutionary model. Bootstrap support values ≥ 60% are indicated at tree nodes. Bar, 5 substitutions per 100 nucleotides.

The first study to apply molecular characterization methods to Lima bean rhizobia in one center of domestication was reported by Ormeño-Orrillo et al. (2006). Twenty one rhizobial strains were isolated from nodules collected in the major cultivated area of this crop in Peru. All isolates were identified as *Bradyrhizobium* based on 16S rRNA and *dnaK* gene sequencing. Two groups were recognized: one could be assigned to the species *Bradyrhizobium yuanmingense* while the other corresponded to a new phylogenetic branch within *Bradyrhizobium* (marked as NB in Figure 1). Genetic heterogeneity within the new lineage was evidenced by sequence analysis of several house-keeping genes (Durán et al., 2014; Ormeño-Orrillo et al., 2006). Two subgroups within this lineage were recently described as the novel species *Bradyrhizobium paxllaeri* and *Bradyrhizobium icense* (Figure 1) (Durán et al., 2014). Bradyrhizobia phenotypically similar to the novel species have been isolated from Lima bean nodules collected in other cultivated areas of Peru (Matsubara Bautista et al., 2005; Matsubara and Zúñiga-Dávila, 2015).

Analysis of symbiotic genes *nodB*, *nodC* and *nifH* revealed that Lima bean bradyrhizobia in Peru did not share a unique symbiotic genotype. Two genotypes were observed in *B. yuanmingense*, while *B. paxllaeri* and *B. icense* each had unique, although related, genotypes. In the *Bradyrhizobium* genus, symbiosis genes are linked and contained in mobile genetic elements known as symbiosis islands (Kaneko et al., 2002). On this basis, it has been speculated that the symbiotic genotypes observed in Lima bean bradyrhizobia could correspond to four distinct symbiosis islands (Ormeño-Orrillo et al., 2006). *B. yuanmingense* strains from Lima bean have nodulation genes belonging to clade III, which is present in pantropical rhizobia (Steenkamp et al., 2008). Nodulation genes of *B. paxllaeri* and *B. icense* belong to clade IV, which is found in rhizobia isolated mostly in subtropical regions from a wide range of legume plants growing in the Mediterranean region, Australia and China (Cardinale et al., 2008; Durán et al., 2013; Ormeño-Orrillo et al., 2013; Stepkowski et al., 2012).

It is now recognized that most species of *Phaseolus* associate preferentially with *Bradyrhizobium* and only a minority, like *P. vulgaris* and *P. coccineus*, prefer to enter in symbiosis with fast growing rhizobial genera like *Rhizobium* or *Sinorhizobium* (Martínez-Romero, 2003; Servín-Garcidueñas et al., 2014). Occasionally, fast-growing rhizobia have been isolated from Lima bean in Peru (Matos-Cuzcano et al., 1998; Ormeño et al., 2007; Zúñiga-Dávila, 2007). Strain LMTR 32, the only strain of the fast-growing rhizobia characterized by molecular methods, was identified as *Sinorhizobium meliloti* and induced effective nodules on Lima bean in laboratory experiments, although it had low nitrogen fixing ability (Ormeño et al., 2007) and tended to lose its symbiotic genes (Ormeño-Orrillo et al., unpublished).

BIODIVERSITY IN THE MESOAMERICAN REGION

Two domestication events in Mesoamerica have been inferred for Lima bean, one occurring in Western central Mexico and the other between Guatemala and Costa Rica (Andueza-Noh et al., 2013). Nowadays, major production areas of Lima bean in Mexico are located in regions bordering the Gulf of Mexico in states like Yucatan and Veracruz.

López-López et al. (2013) isolated Lima bean rhizobia from soils collected in Los Tuxtlas, Veracruz where Mexican farmers grow this crop in association with maize, common bean and cowpea. All ninety eight isolates were ascribed to *Bradyrhizobium* by molecular characterization methods. ERIC-PCR genomic fingerprints revealed that Lima bean bradyrhizobia were more diverse than isolates obtained from cowpea from the same soil samples, probably because cowpea is an exotic crop in the area while Lima bean is native. Seven distinct *Bradyrhizobium* genospecies were present among Lima bean nodule bacteria. Three genospecies, TUXTLAS-33, 34 and 35, were novel while the remaining corresponded to *Bradyrhizobium japonicum*, and *Bradyrhizobium* genospecies TUXTLAS-4, 17 and 22 which were previously reported from forest soils of the same area (Figure 1) (Ormeño-Orrillo et al., 2012). As Lima beans form nodules with bradyrhizobia from tropical forests, they may be well suited for cultivation in other tropical regions. Lima bean nodule bacteria from Los Tuxtlas possessed nodulation genes that belong to clade V, which until now has been found only in rhizobia of American origin.

Wild Lima beans can be found in several regions of Mexico (Andueza-Noh et al., 2013). Two studies have reported the identification of their symbionts based on 16S rRNA gene sequencing. Bacteria from nodules collected in two localities in Oaxaca had sequences similar either to *Bradyrhizobium elkanii* or to *Rhizobium radiobacter* (synonym *Agrobacterium radiobacter*) (Eilmus, 2009). Although closely related to plant pathogenic species, the *R. radiobacter* strain was able to nodulate and fix nitrogen under controlled conditions (Thamer et al., 2011). Four rhizobial nodule isolates from *P. lunatus* obtained in the state of Morelos were identified as *Bradyrhizobium* (Servín-Garcidueñas et al., 2014). Sequence from one isolate grouped within the *B. japonicum* 16S rRNA superclade while the remaining four had sequences which grouped within the *B. elkanii* superclade.

BIODIVERSITY IN OTHER AMERICAN REGIONS

Brazil is not considered as part of the South American center of origin of Lima bean, nevertheless, Northeast Brazilian states are important production areas for this crop (Santos, 2008). The characterization of rhizobia trapped from soils collected in the state of Piaui was reported by Santos et al. (2011). Of fifty isolates obtained, twenty one were fast-growing bacteria while the remaining where classified as intermediate- or slow-growers. Restriction fragment analysis of 16S PCR-amplified genes revealed high diversity among the isolates, which were assigned to *Bradyrhizobium*, *Rhizobium* and *Mesorhizobium*. This was the first time that mesorhizobia were reported as Lima bean nodule bacteria.

Lima beans are also cultivated and widely consumed in the United States (Santamaria, 2007). Based on 16S-23S intergenic spacer sequence analysis, strains USDA 3259 and USDA 3426 obtained in the state of Illinois were assigned to the *B. ekanii* superclade (van Berkum and Fuhrmann, 2009).

CONCLUSION

Despite being described as a restricted host for nodulation (Thies et al., 1991), *P. lunatus* is able to associate with many rhizobial symbionts. At least eleven *Bradyrhizobium* species or genospecies have been found in Lima bean nodules in its centers of domestication (Figure 1). Bradyrhizobia from Andean and Mesoamerican domestication sites are different probably due to differences in their hosts, which were domesticated from distinct gene pools. A higher number of species are reported as being associated with Lima bean crops in Mexico than in Peru. Sites sampled in Peru had monocultures while polycultures with other leguminous plants growing along the Lima bean were sampled in Mexico. The dissimilarities between sites may have contributed to the observed differences in bradyrhizobial species richness between both countries. When summed across countries and sites, most bradyrhizobia associated with the Lima bean belong to the *B. elkanii* 16S rRNA superclade (Figure 1), suggesting a preference to nodulate with species within this lineage. Other *Phaseolus* species do not seem to have this preference (Parker, 2002; Servín-Garcidueñas et al., 2014).

The Lima bean bradyrhizobia from Mexico and Peru are different not only at the species or genospecies level but also in their symbiotic genes. Although this observation may not be unexpected given that *P. lunatus* seems to be a relatively promiscuous host, it raises the question as to whether hosts from one domestication center may preferentially nodulate with rhizobia from the same origin as has been reported for *P. vulgaris* (Aguilar et al., 2004).

Although *Bradyrhizobium* strains are its predominant symbionts, Lima bean is able to effectively nodulate with strains of other rhizobial genera like *Rhizobium*, *Sinorhizobium* and *Mesorhizobium*. It can be speculated that these latter rhizobial genera can gain access to Lima bean nodules in certain edaphic and/or climatic conditions. For example, soybean forms nodules with *Bradyrhizobium* in acid soils and with *Sinorhizobium* in alkaline soils (Tian et al., 2012).

The promiscuity of Lima bean and the high biodiversity of its rhizobial symbionts pose a challenge for programs aimed at using biofertilizers inoculants for this crop, since inoculant strains will have to compete with a relatively high number of other soil rhizobia. Nevertheless, we have observed that certain *Bradyrhizobium* species or genospecies are more common than others in nodules (López-López et al., 2013; Zúñiga-Dávila, 2007) and thus constitute excellent candidates for developing inoculants for Lima bean.

ACKNOWLEDGMENTS

To PAPIIT IN205412 and Concytec-Procyt 229-2005. To M. Dunn for reading the manuscript.

REFERENCES

Aguilar, OM, Riva, O, Peltzer, E. Analysis of *Rhizobium etli* and of its symbiosis with wild *Phaseolus vulgaris* supports coevolution in centers of host diversification. *Proceedings of the National Academy of Sciences of the United States of America*, 2004, 101, 13548-13553.

Allen, ON, Allen, EK. Root nodule bacteria of some tropical leguminous plants: II. Cross-inoculation tests within the cowpea group. *Soil Science*, 1939, 47, 63-76.

Andueza-Noh, RH, Serrano-Serrano, ML, Chacón Sánchez, MI, Sanchéz del Pino, I, Camacho-Pérez, L, Coello-Coello, J, Mijangos Cortes, J, Debouck, DG, Martínez-Castillo, J. Multiple domestications of the Mesoamerican gene pool of lima bean (*Phaseolus lunatus* L.): evidence from chloroplast DNA sequences. *Genetic Resources and Crop Evolution*, 2013, 60, 1069-1086.

Cardinale, M, Lanza, A, Bonnì, M, Marsala, S, Puglia, A, Quatrini, P. Diversity of rhizobia nodulating wild shrubs of Sicily and some neighbouring islands. *Archives of Microbiology*, 2008, 190, 461-470.

Durán, D, Rey, L, Mayo, J, Zúñiga-Dávila, D, Imperial, J, Ruiz-Argüeso, T, Martínez-Romero, E, Ormeño-Orrillo, E. *Bradyrhizobium paxllaeri* sp. nov. and *Bradyrhizobium* icense sp. nov., nitrogen-fixing rhizobial symbionts of Lima bean (*Phaseolus lunatus* L.) in Peru. *International Journal of Systematic and Evolutionary Microbiology*, 2014, 64, 2072-8.

Durán, D, Rey, L, Sánchez-Cañizares, C, Navarro, A, Imperial, J, Ruiz-Argüeso, T. Genetic diversity of indigenous rhizobial symbionts of the *Lupinus mariae-josephae* endemism from alkaline-limed soils within its area of distribution in Eastern Spain. *Systematic and Applied Microbiology*, 2013, 36, 128-36.

Eilmus, S. Diversität und funktionen der mit der ameisengattung Pseudomyrmex (Lund, 1831) assoziierten bakterien. Ph.D. thesis, University of Duisburg-Essen, Leverkusen, 2009, 165 pp.

Gutiérrez Salgado, A, Gepts, P, Debouck, DG. Evidence for two gene pools of the Lima bean, *Phaseolus lunatus* L., in the Americas. *Genetic Resources and Crop Evolution*, 1995, 42, 15-28.

Jordan, DC. Family III. Rhizobiaceae Conn 1938. In: Krieg, NR, Holt, JG (Eds.) *Bergey's Manual of Systematic Bacteriology*. pp. 234–235. 1984. Williams & Wilkins: Baltimore.

Kaneko, T, Nakamura, Y, Sato, S, Minamisawa, K, Uchiumi, T, Sasamoto, S, Watanabe, A, Idesawa, K, Iriguchi, M, Kawashima, K, Kohara, M, Matsumoto, M, Shimpo, S, Tsuruoka, H, Wada, T, Yamada, M, Tabata, S. Complete genomic sequence of nitrogen-fixing symbiotic bacterium *Bradyrhizobium japonicum* USDA110. *DNA Research*, 2002, 9, 189-97.

López-López, A, Negrete-Yankelevich, S, Rogel, MA, Ormeño-Orrillo, E, Martínez, J, Martínez-Romero, E. Native bradyrhizobia from Los Tuxtlas in Mexico are symbionts of *Phaseolus lunatus* (Lima bean). *Systematic and Applied Microbiology*, 2013, 36, 33-38.

Martínez-Romero, E. Diversity of *Rhizobium-Phaseolus vulgaris* symbiosis: overview and perspectives. *Plant and Soil*, 2003, 252, 11-23.

Matos-Cuzcano, G, Ormeño-Orrillo, E, Zúñiga-Dávila, D. Diversidad de los rizobios que nodulan el cultivo de pallar (*Phaseolus lunatus* L.) en la costa central del Perú. *Ecología*, 1998, 1, 42-46.

Matsubara Bautista, M, Ormeño-Orrillo, E, Mayo Hernández, J, Zúñiga Dávila, D. Diversity of rhizobia isolated from *Phaseolus lunatus* L. in the Supe valley of Peru. In: PHASEOMICS IV. Salta, Argentina. 2005, p. 47.

Matsubara, M., Zúñiga-Dávila, D. Phenotypic and molecular differences among rhizobia that nodulate *Phaseolus lunatus* in the Supe valley in Peru. *Ann Microbiol*, 2015, In Press (DOI 10.1007/s13213-015-1054-9).

Ormeño-Orrillo, E, Rogel-Hernández, MA, Lloret, L, López-López, A, Martínez, J, Barois, I, Martínez-Romero, E. Change in land use alters the diversity and composition of *Bradyrhizobium* communities and led to the introduction of *Rhizobium etli* into the tropical rain forest of Los Tuxtlas (Mexico). *Microbial Ecology*, 2012, 63, 822-34.

Ormeño-Orrillo, E, Servín-Garcidueñas, LE, Imperial, J, Rey, L, Ruiz-Argueso, T, Martinez-Romero, E. Phylogenetic evidence of the transfer of *nodZ* and *nolL* genes from *Bradyrhizobium* to other rhizobia. *Molecular Phylogenetics and Evolution*, 2013, 67, 626-630.

Ormeño-Orrillo, E, Vinuesa, P, Zúñiga-Dávila, D, Martínez-Romero, E. Molecular diversity of native bradyrhizobia isolated from Lima bean (*Phaseolus lunatus* L.) in Peru. *Systematic and Applied Microbiology*, 2006, 29, 253-62.

Ormeño, E, Torres, R, Mayo, J, Rivas, R, Peix, A, Velásquez, E, Zúñiga, D. *Phaseolus lunatus* is nodulated by a phosphate solubilizing strain of *Sinorhizobium meliloti* in a Peruvian soil. In: Velázquez, E. Rodríguez-Barrueco, C. (Eds.) First International Meeting on Microbial Phosphate Solubilization. Developments in Plant and Soil Sciences. pp. 143-147. 2007. Springer: Dordrecht.

Parker, MA. Bradyrhizobia from wild *Phaseolus*, *Desmodium*, and *Macroptilium* species in Northern Mexico. *Applied and Environmental Microbiology*, 2002, 68, 2044-2048.

Santamaria, L. Evaluation of lima bean (*Phaseolus lunatus*) germplasm for resistance to downy mildew, and epidemiological and biological studies of its causal agent, *Phytophthora phaseoli*. Ph.D. thesis, University of Delaware, Ann Arbor, 2007, 159 pp.

Santos, JO. Divergência genética em feijão-fava (*Phaseolus lunatus* L.). M.Sc. thesis, Universidade Federal do Piauí, Teresina, 2008, 99 pp.

Santos, JO, Araújo, ASF, Gomes, RLF, Lopes, ACA, Figueiredo, MVB. Rhizobia-*Phaseolus lunatus* symbiosis: importance and diversity in tropical soils - A review. *Dynamic Soil, Dynamic Plant*, 2008, 2, 56-60.

Santos, JO, Antunes, JEL, Araújo, ASF, Lyra, MCCP, Gomes, RLF, Lopes, ACA, Figueiredo, MVB. Genetic diversity among native isolates of rhizobia from *Phaseolus lunatus*. *Annals of Microbiology*, 2011, 61, 437-44.

Servín-Garcidueñas, LE, Zayas-Del Moral, A, Ormeño-Orrillo, E, Rogel, MA, Delgado-Salinas, A, Sánchez, F, Martínez-Romero, E. Symbiont shift towards *Rhizobium* nodulation in a group of phylogenetically related *Phaseolus* species. *Molecular Phylogenetics and Evolution* , 2014, 79, 1-11.

Steenkamp, ET, Stepkowski, T, Przymusiak, A, Botha, WJ, Law, IJ. Cowpea and peanut in southern Africa are nodulated by diverse *Bradyrhizobium* strains harboring nodulation genes that belong to the large pantropical clade common in Africa. *Molecular Phylogenetics and Evolution*, 2008, 48, 1131-44.

Stepkowski, T, Watkin, E, McInnes, A, Gurda, D, Gracz, J, Steenkamp, ET. Distinct *Bradyrhizobium* communities nodulate legumes native to temperate and tropical monsoon Australia. *Molecular Phylogenetics and Evolution*, 2012, 63, 265-277.

Thamer, S, Schädler, M, Bonte, D, Ballhorn, D. Dual benefit from a belowground symbiosis: nitrogen fixing rhizobia promote growth and defense against a specialist herbivore in a cyanogenic plant. *Plant and Soil*, 2011, 341, 209-219.

Thies, JE, Bohlool, BB, Singleton, PW. Subgroups of the cowpea miscellany: symbiotic specificity within *Bradyrhizobium* spp. for *Vigna unguiculata*, *Phaseolus lunatus*, *Arachis hypogaea*, and *Macroptilium atropurpureum*. *Applied and Environmental Microbiology*, 1991, 57, 1540-5.

Tian, CF, Zhou, YJ, Zhang, YM, Li, QQ, Zhang, YZ, Li, DF, Wang, S, Wang, J, Gilbert, LB, Li, YR, Chen, WX. Comparative genomics of rhizobia nodulating soybean suggests extensive recruitment of lineage-specific genes in adaptations. *Proceedings of the National Academy of Sciences of the United States of America*, 2012, 109, 8629-34.

van Berkum, P, Fuhrmann, JJ. Evidence from internally transcribed spacer sequence analysis of soybean strains that extant *Bradyrhizobium* spp. are likely the products of reticulate evolutionary events. *Applied and Environmental Microbiology*, 2009, 75, 78-82.

Zúñiga-Dávila, D. Leguminosas y producción de biofertilizantes en el Perú. In: Izaguirre-Mayoral, ML, Labandera, C, Sanjuán, J (Eds.) Biofertilizantes en Iberoamérica: Una visión técnica, científica y empresarial. pp. 61-67. 2007. BIOFAG-CYTED: Montevideo.

In: Phaseolus Lunatus
Editors: A. S. Ferreira de Araujo et al. ISBN: 978-1-63482-494-1

Chapter 7

RHIZOBIA AND LIMA BEAN SYMBIOSIS: IMPORTANCE, OCCURRENCE, AND DIVERSITY

***Ademir Sérgio Ferreira de Araújo*[*1], *Jadson Emanuel Lopes Antunes*[2], *Ângela Celis de Almeida Lopes*[1], *Regina Lucia Ferreira Gomes*[1] *and Marcia Barreto do Vale Figueiredo*[2]**

[1]Agricultural Science Center, Federal University of Piauí, Teresina, Piauí, Brazil

[2]Soil Biology Laboratory, Agronomical Institute of Pernambuco, Recife, Pernambuco, Brazil

ABSTRACT

The Lima bean (*P. lunatus*) is an important species of *Phaseolus* for human consumption in tropical regions and its seed is an important protein source for populations of South America, Africa and Mexico. In addition, as a legume plant, Lima bean has the ability to perform

[*] Corresponding Author address Dr. Ademir Sergio Ferreira de Araújo. Agricultural Science Center, Federal University of Piauí, Teresina, PI (Brazil); Tel.: +55-86-3215-5740; fax: +55-86-3215-5743. E-mail address: asfaruaj@yahoo.com.br.

biological nitrogen fixation (BNF) due to symbiosis with nitrogen-fixing bacteria. However, the studies about the rhizobia associated with this crop have been scarcely studied. Some studies found Lima bean being nodulated by fast and slow growing rhizobia groups, such as *Bradyrhizobium* and *Rhizobium*. On the other hand, another studies have shown a great diversity of rhizobia nodulating Lima bean where the main genera belongs to *Bradyrhizobium* group. Therefore, more studies are necessary to establish the prevalence of rhizobia species of Lima bean in different regions.

Keywords: Biological nitrogen fixation; legumes; rhizobia diversity

INTRODUCTION

Lima bean (*Phaseolus lunatus*) is an important species of plant for humans in tropical regions. This legume originated in Peru; and archeological evidence supports the hypothesis that lima beans were domesticated in Mesoamerica and South America (Salgado et al. 1995). Lima beans originated from two main groups: one from the lowlands of eastern South America, and another from the western Andes (Salgado et al. 1995). This species is predominantly autogamous, although it has outcrossing rates up to 48% (Baudoin et al. 1998). It is characterized by an indeterminate, climbing growth habit, a prolonged flowering period, and production of a large number of pods (Zoro Bi et al. 2003).

The lima bean seed is considered a main crop and an important source of protein for people from South America, Africa, and Mexico. It also has great relevance in Brazil, mainly in the Northeast, where it is used as an alternative food and as an economic crop. The research unit Embrapa Genetic Resources and Biotechnology has a large collection of germplasm from *P. lunatus*, with approximately 980 subsamples, although it has not been widely studied (Silva 2011). *P. lunatus* seeds vary in size, shape, and color, and they range from flat-seeded types to the round-seeded potato types, which are cream, red, purple, brown, black, and various types of mottled colors (Vieira 1992). The seeds have high protein (210–260 $g \cdot kg^{-1}$) and carbohydrate (550–640 $g \cdot kg^{-1}$) contents, low fat (10–23 $g \cdot kg^{-1}$) levels and high fiber levels (32–68 $g \cdot kg^{-1}$), high levels of the minerals K, Zn, Ca, and Fe, and low levels of Na and P (Chel-Guerrero et al. 2012).

In Brazil, lima bean is the second most economically important species of *Phaseolus* and shows a higher capacity of adaptation (Azevedo et al. 2003).

According to IBGE (2013), Brazil produced approximately 14,951 tons of lima beans from an area of 37,521 ha. The rustic quality of lima bean and its capacity to resist to long, dry periods are important characteristics for the semi-arid region of Northwest Brazil (Azevedo et al. 2003).

The rhizobial microorganisms associated with this crop have been scarcely studied. In the old host-based classification scheme, symbionts of *P. lunatus* were included in the same group as rhizobia associated with slow-growing cowpea (*Vigna unguiculata*). This group was a diverse assemblage of strains that were later included in the genus *Bradyrhizobium* (Quirós et al. 2006). The rhizobial isolates for *P. lunatus* were obtained from areas where this legume is not native, and research has only focused on analysis of symbiotic characteristics, such as infectiveness and effectiveness (Ormenõ-Orrillo et al. 2006).

Biological Nitrogen Fixation in Lima Bean

Lima bean, among the other legumes cultivated throughout the world, has the ability to perform biological nitrogen fixation (BNF) because of a symbiotic relationship with nitrogen-fixing bacteria (Schultze and Kondorosi, 1998).

BNF is an important source of nitrogen (N) for the plant, which is important for plant growth. It is suggested that after photosynthesis, the process performed by plants to produce the energy necessary for their survival, BNF is regarded as the planet's most important biological process (http://cnx.org/content/m47338/latest/). This process occurs as a symbiotic system between legumes and rhizobia (Zahran, 2001), and the fixed N_2 represents a renewable source of N within agriculture. The total quantity of N fixed by plants is estimated to range from 100 to 900 kg $ha^{-1} \cdot year^{-1}$.

In agricultural systems, approximately 80% of the fixed N_2 originates from symbioses between legumes and the nitrogen-fixing bacteria. Total global BNF derived from rhizobium–legume symbiosis is estimated to be 50–70 Tg $N \cdot year^{-1}$, which is comparable with the total amount of N from fertilizers, 80–90 Tg $N \cdot year^{-1}$ (Herridge et al. 2008).

The use of microorganisms that enhance nodulation and BNF is of fundamental importance because it may contribute to increased plant productivity. The microorganisms beneficial to plants are of interest to the agricultural sector in view of their ecologically beneficial effects (Compant and Sessitsch 2010). Bacteria capable of producing nodules and fixing N_2 in legumes are divided into more than 50 species that are distributed among the genera *Rhizobium*, *Azorhizobium*, *Bradyrhizobium*, *Burkholderia*, *Cupriavidus*, *Devosia*, *Ensifer*, *Mesorhizobium*, *Methylobacterium*, *Ochrobactrum*, *Phyllobacterium*, *Shinella*, *Pseudomonas*, and *Microvirga* (Shiraishiet al. 2010; Ardley et al. 2012; Weir 2012).

The success of symbiosis depends on the availability of compatible rhizobia for the specific legumes. A great number of legumes were introduced into new areas with inoculants containing the nitrogen-fixing bacteria (Mutch and Young 2004). The use of inoculants for soybean in Brazil saves the economy approximately 7 billion dollars per year and these crops cover an area of approximately 24 million hectares (Alves et al. 2003; Embrapa 2012). However, studies on BNF in soybean crops are already consolidated (Hungria et al. 1998; Peoples and Herridge 2000; Campo et al. 2009; Zilli et al. 2010; Torres et al. 2012).

Considering the genus *Phaseolus*, the published studies have presented information about nodulation and nitrogen fixation in the common bean (*P. vulgaris*), which forms symbiotic associations with several species from the genus *Rhizobium* (Martinez-Romero et al. 1991; Souza et al. 1994; Martinez-Romero 2003). Therefore, there is adequate existing knowledge regarding the interactions between rhizobia and common bean, nodule development, and the process of nitrogen fixation. Previously, the symbiotic process in lima bean has been studied using bacteria from the subgroups of the cowpea from various *Bradyrhizobium* sp. (Allen and Allen 1939; Thies et al. 1991; Thies and Bohlool 1991). The results show that the lima bean has the ability to nodulate with *Bradyrhizobium* sp.

To investigate nodule development in lima bean, Santos et al. (2009) evaluated the ontogeny of nodulation, and they found that the highest nodule number and mass occurred between 45 and 60 days after plant emergence (Table 1), which corresponded to the flowering period of the plant. For common bean, the period of the highest nodule number and mass occurs 30 days after plant emergence (Hungria and Neves 1986).

Table 1. Nodulation in Lima bean at 30, 45 and 60 days after plant emergence (Santos et al. 2010)

Day	Nodule number (nod pl^{-1})	Nodule mass (g pl^{-1})	Nodule size (mg nod^{-1})
30	188 b	1.03 b	50 b
45	371 a	2.40 a	70 ab
60	367 a	3.37 a	90 a

Additionally, Santos et al. (2009) found red colored nodules as an indication of presence of leghemoglobin (Figure 1). Leghemoglobin provides an O_2 flux for rhizobial respiration, and maintain O_2 at a concentration that does not render the nitrogenase complex inactive (Appleby 1984). Red-colored nodules also indicate that BNF is occurring satisfactorily. Conversely, evaluation of the efficiency of N_2 fixation is necessary to indicate high nodule activity. Antunes et al. (2011) evaluated the symbiotic effectiveness of 17 rhizobial isolates from lima bean in Northeast Brazil, and they compared the isolates with two reference *Rhizobium* strains CIAT 899 and NGR 234. They found eight isolates with higher N accumulation and N_2-fixation efficiency compared with the reference strains CIAT 899 and NGR 234. They also evaluated the morphological and biochemical characteristics of these isolates, and they found six isolates with characteristics of *Bradyrhizobium* and two isolates with characteristics of *Rhizobium.*

Figure 1. Nodules in Lima bean roots at 45 days after plant emergence. (A) Nodules in the roots; (B) separated nodules; (C) Leg-hemoglobin pigment in nodule of Lima bean (Source: Jadson E. L. Antunes).

Diversity of Rhizobia That Promote Nodulation in Lima Bean

Rhizobia are common inhabitants of diverse soil types, and population sizes can substantially differ, from a few cells to 10^5 bacteria per gram of soil

(Drew and Ballard 2010). The wide geographical distribution of rhizobia may be related to the large genetic diversity among these bacteria (Martinez-Romero and Caballero-Mellado 1996). Currently, studies have been conducted with molecular methods that provide the possibility to identify taxonomic groups of rhizobia in soils (Santos et al. 2008). Considering the genus *Phaseolus*, the common bean (*P. vulgaris*) has been found to be nodulated by *Rhizobium*, which belongs to fast-growing groups of rhizobia (Oliveira et al. 2010). Lima bean are nodulated by fast and slow growing groups of rhizobia (Santos et al. 2011), and some studies found *Bradyrhizobium* and *Rhizobium* as symbionts of this legume (Thies et al. 1991; Thies and Bohlool 1991; Santos et al. 2011). The few studies of the genetic diversity among rhizobia from lima bean have examined a relative wide collection from geographic locations including Peru, Mexico, and Brazil (Ormeno-Orrillo et al. 2006, 2007; Lopez et al. 2013; Santos et al. 2011; Lopes et al. 2013). Because Peru is the center of diversity for lima bean; therefore, some studies were conducted there with the aim to characterize the diversity of rhizobial symbionts of lima bean (Ormeno-Orrillo et al. 2006; 2007). Ormeño-Orrillo et al. (2006) evaluated the molecular diversity of rhizobial isolates associated with lima bean in Peru. They found divergent bradyrhizobial lineages (*Bradyrhizobium yuanmingense* and *Bradyrhizobium* sp.) according to PCR-RFLP of the *rpoB* gene and sequence analysis of the 16S rDNA and *dnaK*, *nifH*, and *nodB* genes. According to Ormeño-Orrillo et al. (2006), bradyrhizobia is the main species effectively nodulating lima bean in Peru.

Recently, Lopes et al. (2013) described the nodule bacteria from native lima beans from Mexico. The bacterial diversity of isolates from nitrogen-fixing nodules of *P. lunatus*, using ERIC-PCR and PCR-RFLP of *rpoB* genes and sequencing of *recA*, *nodZ*, and *nifH* genes, show that nodule bacteria correspond to *Bradyrhizobium*. According to Lopes et al. (2013), this is the first report of nodule bacteria from *P. lunatus* in its Mesoamerican site of origin and domestication and it may confirm that *Bradyrhizobium* is the main nodulating group of lima bean. However, the fast growing rhizobia that nodulate *Phaseolus* commonly belong to the genus *Rhizobium* and *Sinorhizobium*, and Ormeno et al. (2007) found a strain of *S. meliloti* isolated from lima bean in Peru, and they suggested that this rhizobial species may also nodulate this legume.

A study about the genetic diversity of native rhizobia that nodulate lima bean from Brazil shows a broad spectrum of rhizobial groups associated with lima bean (Santos et al. 2011). In that study, rhizobia isolates were obtained and placed into groups based on divergence from the morphological, physiological, and genetic characterization. The restriction patterns obtained with endonucleases MboI, HaeIII, and NheI showed sufficient variability to discriminate isolates identified as species from the genera *Bradyrhizobium*, *Mesorhizobium*, and *Rhizobium*. Antunes (2014 unpublished data) sequenced the 16S rDNA of the above isolates and found that species that nodulate lima bean belonged to the genus *Bradyrhizobium*, *Sinorhizobium*, and *Rhizobium*. These results confirm that lima bean may be nodulated by diverse rhizobia species.

CONCLUSION

The Lima bean is an important legume to tropical regions, and the published studies show a great diversity of rhizobia that nodulate lima bean, although they mainly belong to the genera *Bradyrhizobium*. However, more studies are necessary to establish the prevalence of different rhizobial species of lima bean in different regions, in addition to the urgent need for the improved productivity and NBF efficiency of this crop.

ACKNOWLEDGMENTS

The authors are grateful to 'Conselho Nacional de Desenvolvimento Científico e Tecnológico' (CNPq-Brazil) for financial support of research with Lima bean. All authors are supported by Research fellowships from CNPq-Brazil (Research Productivity).

REFERENCES

Alves BJR, Boddey RM, Urquiaga S (2003) The success of BNF in soybean in Brazil. *Plant Soil* 252:1-9.

Antunes JEL, Gomes RLF, Lopes ACA, Araújo ASF, Lyra MCCP, Figueiredo MVB (2011) Eficiência simbiótica de isolados de rizóbio noduladores de feijão-fava (Phaseolus luntaus L.). *Rev. Bras. Ci. Solo* 35:751-757.

Appleby CA (1984) Leghemoglobin and Rhizobium respiration. *Ann. Rev. Plant Physiol.* 35:521-554.

Ardley JK, Parker MA, Meyer SE, Trengove RD, O'Hara GW, Reeve WG, Howieson JG (2012) Microvirga lupinisp. nov., Microvirga lotononidis sp. nov., and Microvirga zambiensis sp. nov. are Alphaproteobacterial root nodule bacteria that's pecifically nodulate and fix nitrogen with geographically and taxonomically separate legume hosts. *Int. J. Syst. Evol. Microbiol.* 62:2579-2588.

Azevedo JN, Franco JD, Araújo ROC (2003) Composição química de sete variedades de feijão-fava. Teresina, Embrapa Meio-Norte, 4 pp.

Baudoin JP, Degreef J, Hardy O, Janart F, Zoro Bi I (1998) Development ofan in situ conservation strategy for wild Lima bean (Phaseolus lunatus L.) populations in the central valley of Costa Rica. In: Owens SJ, Rudall PJ (Eds) Reproduction Biology, Royal Botanic Garden Press, England, pp 417-426.

Compant SS, Clément C, Sessitsch A (2010) Plant growth-promoting bacteria in the rhizo- and endosphere of plants: their role, colonization, mechanisms involved and prospects for utilization. *Soil Biol. Biochem.* 42:669-678.

Campo RJ, Araujo RS, Hungria M (2009) Nitrogen fixation with the soybean crop in Brazil: compatibility between seed treatment with fungicides and bradyrhizobial inoculants. *Symbiosis* 48:154-163.

Chel-Guerrero L, Domínguez-Magaña M, Martínez-Ayala A, Dávila-Ortiz G, Betancur-Ancona D (2012) Lima Bean (Phaseolus lunatus) Protein Hydrolysates with ACE-I Inhibitory Activity. *Food Nutrit Sci.* 3:511-521.

Embrapa (2012) XXI Science for life. Ministry of Agriculture, Livestock and Food Supply. Brazil, 22p.

Hungria M, Neves MCP (1986) Ontogenia da fixação biológica do nitrogênio em Phaseolus vulgaris. *Pesq. Agropec. Bras.* 21:715-730.

Hungria M, Boddey LH, Santos MA, Vargas MAT (1998) Nitro-gen fixation capacity and nodule occupancy by Bradyrhizobium japonicum and B. elkanii strains. *Biol. Fertil Soils* 27:393–399.

Martínez-Romero E (2003) Diversity of Rhizobium-Phaseolus vulgaris symbiosis: overview and perspectives. *Plant and Soil* 252:11-23.

Ormenõ-Orrillo E, Vinuesa P, Zuñiga-Dávila D, Martínez-Romero E (2006). Molecular diversity of native bradyrhizobia isolated from Lima bean (Phaseolus lunatus L.) in Peru. *Syst. App. Microbiol.* 29: 253–262.

Ormeño E, Torres R, Mayo J, Rivas R, Peix A, Velázquez E, Zúniga D (2007) Phaseolus lunatus is nodulated by a phosphate solubilizing strain of Sinorhizobium meliloti in a Peruvian soil. In: Velázquez E, Rodríguez-

Barrueco C (Eds) Developments in Plant and Soil Sciences, Springer-Verlag, The Netherlands, pp 243-247.

Ormeño-Orrillo E, Vinuesa P, Zuniga-Davila D, Martínez-Romero E (2006) Molecular diversity of native bradyrhizobia isolated from (Phaseolus lunatus L.) in Peru. *Syst. App. Microbiol.* 29:253-226.

Peoples MB, Herridge DF (2000) Quantification of biological nitrogen fixation in agricultural systems. In Nitrogen Fixation: From Molecules to Crop Productivity. Pedrosa F (Ed). Kluwer Academic Publishers, Dordrecht, 519-524.

Quirós A, Chichón R, Recio I, López-Fandiño R (2006) The Use of High Hydrostatic Pressure to Promote the Proteolysis and Release of Bioactive Peptides from Ovalbumin. *Food Chem.* 54:19-27.

Salgado AG, Gepts P, Debouck DG (1995) Evidence for two gene pools of the Lima bean, Phaseolus lunatus L., in the Americas. *Gen Res. Crop Evol.* 42:15-28.

Santos JO, Araújo ASF, Gomes RLF, Lopes ACA, Figueiredo MVB (2009) Ontogeny of nodulation in Lima bean (Phaseolus lunatus). *Braz J. Agric. Res.* 4:426-429.

Santos JO, Antunes JEL, Lyra MCCP, Araujo ASF, Gomes RLF, Lopes ACA, Figueiredo MVB (2011) Genetic diversity among isolates of Rhizobia from Phaseolus lunatus. *Annals Microbiol.* 61:437-444.

Schultze M, Kondorosi A (1998) Regulation of symbiotic root nodule development. *Ann. Rev. Genet.* 32:33-57.

Shiraishi A, Matsushita N, Hougetsu T (2010) Nodulation in blacklocust by the Gammaproteobacteria Pseudomonas sp. and the Betaproteobacteria Burkholderia sp. *Sys. App. Microbiol.* 33:269–274.

Silva RNO (2011) Diversidade genética em feijão-fava (Phaseolus lunatus L.) por marcadores morfoagronômicos e moleculares. *Dissertation of Master*, UFPI, 2011, p.175

Silva JA, Oliveira MG, Souza LT, Assunção IP, Lima GA, Michereff SJ (2014). Reação de genótipos de feijão-fava a Sclerotium rolfsii. *Hortic Bras.* 32:98-101.

Souza V, Eguiarte L, Ávila G, Capello R, Gallardo C, Montoya J, Pinero D (1994) Genetic structure of Rhizobium etli biovar phaseoli associated with wild and cultivated bean plants (Phaseolus vulgaris and Phaseolus coccineus) in Morelos, Mexico. *App. Environ. Microbiol.* 60:1260-1268

Torres AR, Kaschuk G, Saridakis GP, Hungria M (2012) Genetic variability in Bradyrhizobium japonicum strains nodulating soybean [Glycine max (L.) Merrill]. *World J. Microbiol. Biotechnol.* 28: 1831-1835.

Vargas MAT, Peres JRR, Suhet AR (1982) Adubação nitrogenada, inoculação e épocas de calagem para a soja em um solo do cerrado. *Pesq. Agropec. Bras.* 17:1127-1132.

Vieira RFA (1992) Cultura do feijão-fava. *Inf. Agropec.* 16:30-37

Xavier TF, Araújo ASF, Santos VB, Campos FL (2007) Ontogenia da nodulação em duas cultivares de feijão-caupi. *Ci. Rural* 37:572-575

Weir BS (2012) The current taxonomy of rhizobia. New Zealand rhizobia website. http://www.rhizobia.co.nz/taxonomy/rhizobia.html. Last updated: 14 September. Acessado em 02/04/2012.

Zilli JE, Valicheski RR, Rumjanek NG, Araújo JLS, Freire Filho FR, Neves MCP (2006) Eficiência simbiótica de estirpes de Bradyrhizobium isoladas do solo de cerrado em caupi. *Pesq. Agropec. Bras.* 41:811–818.

Zilli JE, Campo RJ, Hungria M (2010) Eficácia da inoculação de Bradyrhizobium em pré-semeadura da soja. *Pesq. Agropec. Bras.* 45:335-338.

Zoro Bi I, Maquet A, Baudoin JP (2003) Population genetic structure of wild Phaseolus lunatus (Fabaceae), with special reference to population sizes. *Am. J. Bot.* 90:897-904.

In: Phaseolus Lunatus
Editors: A. S. Ferreira de Araujo et al.
ISBN: 978-1-63482-494-1

Chapter 8

LIMA BEAN DISEASES

Eulália Maria Sousa Carvalho*[*1], José Evando Aguiar Beserra Jr.[1] and Beatriz Meireles Barguil[2]

[1]Universidade Federal do Piauí, Centro de Ciências Agrárias, Departamento de Fitotecnia, Teresina-PI, Brazil
[2]Universidade Estadual do Piauí, Centro de Ciências da Natureza, Teresina-PI, Brazil

ABSTRACT

Plant diseases are disorders caused by the continuous action of infectious agents that change plant metabolism when they infect plants or their organs, compromising productivity and/or product quality. Plant diseases, including those in lima bean, are primarily caused by fungi, viruses, nematodes and bacteria. Notably, anthracnose, web blight, viral diseases and root-knot disease are the primary main lima bean diseases in Brazil. The main diseases of this crop in Brazil and worldwide will be introduced in this chapter, highlighting their etiologic agents, symptoms, epidemiology and control whenever possible. Because of the scarcity of studies on lima bean diseases, at times, symptoms and control measures

* Corresponding Author address Dr. Eulalia M. S. Carvalho. Agricultural Science Center, Federal University of Piauí, Teresina, PI (Brazil); Tel.: +55-86-3215-5740; fax: +55-86-3215-5743. E-mail address: eulaliamsc@yahoo.com.br

used in other crops of the same botanical family, such as the common bean, will be introduced.

INTRODUCTION

Lima bean (*Phaseolus lunatus* L.) is an important source of food for the Brazilian population, especially in the Northeast region, because of its hardiness and the high protein content of its grains (Chel-Guerrero et al., 2011).

In 2013, the area planted with lima bean in the State of Piauí was 2,204 hectares (ha), with a production of 266 tonnes (t) and a mean yield of 162 kg/ha (IBGE, 2014) in intercropping primarily with maize on small farms (Silva et al., 2010). Low crop yield results from several factors, including the failure to adopt production technology (Santos et al., 2002), the lack of mineral and organic fertilisation programmes (Alves et al., 2008) and the occurrence of pests and diseases (Vieira, 1992).

Several diseases occur in lima beans, including viral diseases (Sengooba et al., 1997; Melgarejo et al., 2007; Ramos-Sobrinho et al., 2014), anthracnose and web blight (Paula Júnior et al., 1995), mildew (Evans et al., 2002; 2007; Davidson et al., 2008), rust (LYNCH et al., 2006; Bonde et al., 2008; Slaminko et al., 2008) and pod necrosis (Davidson et al., 2002; DAVEY et al., 2008).

The lack of disease records in lima bean cultivation contributes to the ignorance of the actual prevalence of diseases in a specific region and the impact they have on crop yield.

The diseases presented below were discovered in literature searches, where at times, *P. lunatus* was cited only as the host plant of a specific pathogen. However, these diseases may be important for the crop in the future. Where there was a lack of information, symptoms and control measures used in other crops of the same botanical family are discussed.

VIRAL DISEASES

In Brazil, viral diseases have been cited as the most important diseases of lime bean; however, there are few reports regarding these diseases (Vieira, 1992; Costa, 1995; Ramos-Sobrinho et al., 2014).

Golden Mosaic Disease

The bean golden mosaic disease caused by the bean golden mosaic virus (BGMV), family *Geminiviridae*, genus *Begomovirus*, is one of the main diseases of lima bean; this disease has complicated or even precluded bean production in several Brazilian regions (Aragão et al., 2013). The occurrence of this disease in the Northeast region is widespread with an incidence of up to 100%, especially during the flowering and pod formation stages, which coincide with decreased rainfall and, therefore, increased insect vector population (Costa, 1995).

Symptoms manifest as a bright yellow colour that is noticeable throughout plant shoots, which allows the identification of its occurrence from a distance in plantations. Leaf size decreases without morphological changes, and the plant crown volume decreases (Costa, 1995). Transmission occurs through the silverleaf whitefly *Bemisia tabaci* (Genn.) (Hemiptera: Aleyrodidae), which may acquire the virus from other legume crops, including soybean (*Glycine max* L.) and common bean (*Phaseolus vulgaris* L.; Ramos-Sobrinho et al., 2014).

BGMV is the only begomovirus capable of infecting lima bean plants in Brazil. However, soybean plants are known to serve as an inoculum source of other begomoviruses for lima bean plants, as three begomovirus strains were reported to be naturally infecting soybean plants in the central region of Brazil (Fernandes et al., 2009; Coco et al., 2013). BGMV is not transmitted by seeds or mechanical inoculation with infected plant extracts (Costa, 1995).

The genetic diversity analysis of BGMV collected from lima bean plants in the states of Alagoas and Pernambuco, Northeast Brazil, showed identity rates of up to 94% with a soybean BGMV isolate. Furthermore, all of the isolates showed phylogenetic relationships with the soybean and common bean BGMV (Ramos-Sobrinho et al., 2014). The expansion of vector host species cultivation, including soybean, common bean and cowpea (*Vigna unguiculata* (L.) Walp. subsp. *unguiculata*), may cause rapid spread of the virus.

No control measures have been established for golden mosaic disease in lima bean. However, control measures for common bean have primarily focused on controlling the insect vector with contact or systemic insecticides. The use of partly resistant or tolerant varieties is another control measure used, although no lima bean varieties with an adequate level of resistance to golden mosaic are found in Brazil. Therefore, the control of golden mosaic disease also depends on farming practices that may facilitate disease management, including the choice of planting season, avoiding the overlap between the start of cultivation and high populations of silverleaf whiteflies, and intercropping or cropping between tracks with other species used as barriers to the vector (Lemos et al., 2003). The use of a drip irrigation system is also known to be four times more effective than sprinkling at reducing the incidence of golden mosaic disease in common bean plants (Lima et al., 2011).

Mosaic Diseases

In Brazil, at least three viral strains are associated with mosaic disease symptoms in lima bean. The symptoms of mosaic disease begin with a whitening of the first trifoliate leaf veins, followed by a weak mosaic, with mild leaf deformation and smaller leaf size compared to healthy leaves. Plant height is affected, although there are no data on the effect of this disease on crop yield (Costa, 1995).

Using mechanical inoculation, Costa (1983) noted the susceptibility of three lima bean varieties to cowpea severe mosaic virus (CPSMV), family *Secoviridae*, genus *Comovirus*. The plants showed symptoms of diffuse chlorotic lesions, followed by mosaic and systemic vein necrosis. Subsequently, a comovirus, possibly CPSMV, was identified based on the biological, morphological and serological characteristics of lima bean seedlings resulting from seed germination, confirming the viral transmission through seeds; the transmission frequency is unknown (Costa et al., 1991a).

The same authors also identified cowpea aphid-borne mosaic virus (CABMV), family *Potyviridae*, genus *Potyvirus*, in naturally infected lima bean plants by transmission electron microscopy, host range and serology (Costa et al., 1991b). Subsequently, CABMV was identified in lima bean plants with mosaic symptoms in different producing regions of the State of Pernambuco. Mosaic disease symptoms are frequently observed in lima bean, especially in intercropping systems with cowpea (Andrade et al., 2001).

Another virus-causing mosaic found in lima bean is Cucumber mosaic virus (CMV), family *Bromoviridae*, genus *Cucumovirus*. Natural infection was confirmed serologically (Lima; Santos, 1985). Cowpea plants grown near lima bean fields showed mixed infection with CABMV and exhibited severe mosaic symptoms and systemic necrosis.

The virus is transmitted through different aphid species in the field and mechanically in greenhouses (Costa, 1995). Studies of seed transmission of viruses are scarce, although Costa et al. (1991a) observed CPSMV transmission. There are no control measures specific for viruses causing mosaic disease in lima bean.

Other Viral Diseases

Southern bean mosaic virus (SBMV), genus *Sobemovirus*, a strain naturally found in common bean, was capable of infecting lima bean by mechanical inoculation, although its occurrence in plants grown in the field has not yet been reported (Costa, 1983).

The common mosaic disease, a disease commonly found in common bean, may also occur in lima bean, at least in Peru (Melgarejo et al., 2007). The disease is caused by bean common mosaic virus (BCMV), family *Potyviridae*, genus *Potyvirus*. In Brazil, before the appearance of resistant varieties, the common mosaic disease was the most economically important viral disease for common bean because of the damage it caused and its seed transmission (Bianchini et al., 2005). Thus, this disease may also become important for lima bean. The most common symptoms in common bean are leaf mosaic along the leaf veins or formation of bubbles in the dark green areas, curling, and wringing and decreased size of leaflets when the symptoms are more severe.

Seed transmission in common bean ranges from 3% to 95%, depending on the plant variety, viral strain and plant stage at infection. Transmission in plants may occur through mechanical inoculation or aphids, with *Myzus persicae* and *M. nicotianae* among the most effective species. The use of varieties with gene *I* has been effective in controlling this viral disease. In cases when these varieties are not available, the use of virus-free seeds, combined with the use of insecticides to control aphids, is an alternative that may also prevent common mosaic disease.

In Peru, bean common mosaic necrosis virus (BCMNV), family *Potyviridae*, genus *Potyvirus*, was also found in lima bean (Melgarejo et al., 2007). There are no records of occurrence in Brazil, which does not necessarily mean that this viral disease is absent from producing regions.

In addition to BGMV, another begomovirus, Mungbean yellow mosaic India virus (MYMIV), is capable of infecting and inducing yellow mosaic symptoms in lima bean. In India and Nepal, this virus is highly prevalent and is considered an important pathogen of bean plants (Shahid et al., 2012).

FUNGAL DISEASES

Anthracnose

Although no loss estimates have been published, anthracnose caused by the fungus *Colletotrichum truncatum* (Schw.) Andrus and Moore is commonly found in lima bean plantations, especially those in Northeast Brazil. In Piauí, the disease is found in producing areas. Its occurrence has been causing serious problems, with considerable damage to the crop in the experimental fields in the Centre for Agricultural Sciences of the Federal University of Piauí (Universidade Federal do Piauí - UFPI).

The disease is characterised by reddish spots that are initially observed along the leaf veins, on the underside of leaves and on young stems and petioles (Figure 1). The pod lesions are depressed, large and reddish; the pathogen acervuli, which have a milky mass appearance, form on the lesions, from where numerous spores emerge (Paula Júnior et al., 1995). Red and necrotic lesions may also be observed in the hypocotyls, stems, leaf veins and leaf surface, resulting in wrinkled leaves and underdeveloped plants (Costa, 1986). According to Carvalho (2009), the first symptoms appeared three days after fungal inoculation of plants grown in the greenhouse, with the occurrence or not of falling leaves or of one or more leaflets.

Regarding environmental conditions, Pereira (2014) found that 48 hours of leaf wetness favoured disease severity more strongly than 24 hours of leaf wetness and that temperatures ranging from 24 to 27°C were more adequate than 33°C for disease development. The author also noted that leaf age had no effect on susceptibility to the pathogen.

Studies regarding resistance to anthracnose in lima beans are scarce. In studies conducted by Cavalcante et al. (2012), only two accessions (UFPI 251 and UFPI 673) from a total of 30 tested accessions from the Lima Bean Germplasm Bank of the UFPI were highly resistant to *C. truncatum*. In soybean and lentils, there are reports of variability in genotypic resistance to the disease and genetic diversity of *C. truncatum* isolates (Nechet et al., 2004; Buchwaldt et al., 2004).

The use of farming practices, including crop rotation with grasses and use of resistant varieties, healthy seeds and chemicals, has been recommended to control common bean anthracnose (Bianchini et al., 2005). The use of healthy seeds and fungicide application have been the control measures used because there are no reports of resistant varieties of lima bean, although no pesticides specific for the crop have been registered. In common bean, fungicide application is recommended when the first disease symptoms are observed.

Figure 1. Fungal diseases in lima bean. A-Anthracnose symptoms leaf. B- Anthracnose symptoms pod. C-*Colletotrichum truncatum* acervuli. D-Web blight symptoms. Teresina-Piauí, Brazil.

Web Blight

This disease is caused by the fungus *Thanatephorus cucumeris* (Frank) Donk. (*Rhizoctonia solani* Kühn), which usually occurs under conditions of high temperature, frequent rainfall and high relative humidity. It is particularly important in lima bean plantations in the Northeast region. The disease may cause losses of up to 100%, 80% and 60% in the production of common bean, cowpea and soybean, respectively (Paula Júnior et al., 1995; Nechet et al.,

2004; Nechet; Halfeld Vieira, 2006). No loss estimates have been reported for lima bean.

The initial symptoms of web blight are small, wet, brownish, circular lesions that are first observed in the leaves closest to the soil. As the disease progresses, the lesions acquire an irregular shape, are lighter in the centre and are outlined by a dark edge. The aerial mycelium of the fungus is then formed, with a light brown web appearance, connecting an affected leaf to another leaf and forming a "sticky" bond (Figure 1). Under favourable climatic conditions, the mycelial web reaches adjacent leaves, also connecting the petiole, stems, flowers, pods, and neighbouring plant shoots. The pod lesions are dark and depressed, and the affected seeds are malformed. The dry leaves usually remain adhered to the stems by the fungal mycelium. A large number of small, cream-coloured microsclerotia with ill-defined shape are formed that act as pathogen vectors. Small, circular, reddish-brown lesions caused by basidiospores may be observed on the leaves (Paula Júnior et al., 1995).

Because of the close contact between neighbouring plants, the disease quickly spreads through planting areas; it also spreads through raindrops carrying pathogen basidiospores from the soil to leaves. Due to the dissemination and survival characteristics of the causal agent, the adoption of several control measures is essential for disease control. The use of healthy seeds, disposal of infected crop residues and rotation with non-host crops, including tobacco and grasses, are some of the farming practices recommended. Sprinkler irrigation systems must be controlled to avoid excessive soil moisture. Fungicide application is recommended if performed in the early stages of the disease (Paula Júnior et al., 1995). Some active ingredients have been registered with the National Health Surveillance Agency (Agência Nacional de Vigilância Sanitária – ANVISA) for use in soybean stages susceptible to web blight (Nechet et al., 2004).

Lima bean genetic resistance to web blight is still unavailable to Brazilian producers. However, Assunção et al. (2011) identified four genotypes (F-42, F-49, F-53 and F-58) that are highly resistant to the disease from a total of 72 genotypes tested. No genotype showed an immunity reaction to *R. solani.*

Stem Rot Disease

Stem rot disease is caused by the fungus *Sclerotium rolfsii* Sacc., which may cause rot in the plant crown area, with underlying symptoms of progressive yellowing and defoliation in the upper branches and sudden wilting leading to total drying. It favours high soil moisture.

The control of stem rot disease in lima bean is very difficult. Control measures, including the use of well-drained soils and moderate irrigation as well as crop rotation with grasses, may help reduce the pathogen population. There are no pathogen-resistant varieties available to producers. Studies on the reaction of lima bean genotypes to *S. rolfsii* identified two highly resistant genotypes (F-2 and F-25) in a population of 50 tested genotypes (Silva et al., 2014).

Mildew

In lima bean, the mildew caused by the fungus *Phytophthora phaseoli* Thaxter is a destructive disease that is considered to be one of the most important plant diseases in the eastern United States (Evans et al., 2007).

The symptoms are more severe in pods, although they also occur in branches, leaves and petioles. Early leaf lesions are irregular and purple; a white felt consisting of mycelium develops in pods, and pathogen sporangia may cover the entire leaf surface. Reddish bands usually outline the infected areas in pods, which wilt and become unsuitable for consumption. The pathogen requires high humidity for infection. The sporangia are spread by air and insects and may also be carried by slugs while feeding on leaves and cotyledons. The pathogen survives in the soil as oospores (Bezerra et al., 2001).

The pathogen shows high genetic variability. Six *P. phaseoli* physiological strains are known (A, B, C, D, E and F). The use of resistant varieties is, therefore, the most efficient disease control method (Evans et al., 2007; Davidson et al., 2008). Varieties resistant to strains A, B and C have been reported. Chemical control using copper-based fungicides, maneb, metalaxyl and fosetyl-Al is also effective (Erwin; Ribeiro, 1996 apud Bezerra et al., 2001).

Rust

Rust is a disease that may be caused by three fungal species: *Uromyces phaseoli* var. *typica, Phakopsora pachyrhizi* Syd. & P. Syd. and *Phakopsora meibomiae* (Figueiredo, 2001; Bianchini et al., 2005; Lynch et al., 2006). Rust caused by *U. phaseoli* var. *typica* is one of the most important lima bean diseases in Africa (Vieira, et al., 1992). In Brazil, this disease is most common in *Phaseolus vulgaris* grown in humid tropical and subtropical regions. However, the pathogen may also infect other species of *Phaseolus*, in addition to *Vigna* spp. The symptoms appear in pods and stems, albeit predominantly in leaves, and are visible six days after infection as small, whitish and slightly protruding spots. Those spots increase in diameter and become yellow, and the lesions rupture at 7-9 days, forming reddish-brown pustules. The typical symptoms are observed approximately 10-12 days after infection and are characterised by pustules 1-2 mm in diameter and abundant production of urediniospores, commonly surrounded by a chlorotic halo. The urediniospores are spread by humans, farm equipment, animals and especially wind (Bianchini et al., 2005).

Crop rotation, removal of crop residues and disposal of volunteer host plants are recommended for disease control to reduce the initial inoculum. Rust control through genetic resistance is hindered due to the wide pathogenic variability of the causal agent. Some common bean varieties show resistance. There are no reports of resistant varieties for lima bean. Chemical control should be performed at the onset of the first pustules in areas with high incidence. The species *P. meibomiae,* the causal agent of "American rust", also occurs naturally in a variety of legumes, including lima bean (Figueiredo, 2001). It rarely causes economic losses in soybean.

In contrast, "Asian rust" caused by *P. pachyrhizi* may be found in virtually all Brazilian soybean-producing areas and can cause up to a 75% reduction in productivity (Bianchini et al., 2005). The disease is favoured by well-distributed rainfall and long periods of wetness. Lima bean rust caused by *P. pachyrhizi* has been found in the United States (Lynch et al., 2006). There are no reports of its occurrence in Brazil, although it may cause damage to the crop if planted near soybean plantations.

Asian rust symptoms in soybean usually occur starting in the plant reproductive stage. The first symptoms are characterised by tiny spots (maximum of 1 mm in diameter) that are darker than the healthy leaf tissue, with greenish to greenish-grey colour and a corresponding protuberance (uredium) on the leaf underside. The uredia become light brown to dark brown and open into a tiny pore, releasing the hyaline spores that accumulate around the pores, which are carried by the wind. The leaf tissue around the uredia becomes light brown to reddish-brown. Infection with *P. pachyrhizi* causes rapid yellowing or browning and premature leaf fall. The earlier the defoliation occurs, the smaller the grain size will be, and therefore, the greater the loss of green bean yield and quality will be (Embrapa Soja, 2008).

The existence of multiple strains complicates disease control through varietal resistance, and chemical control is the most viable measure to prevent losses caused by rust.

Root Rot and Wilt Disease

Several pathogens reportedly cause root rot and wilt disease in lima bean; the most important species are *Rhizoctonia solani*, *Fusarium solani* f. sp. *phaseoli*, *Pythium ultimum* and *Thielaviopsis* (Vieira, 1992). *Phytophthora capsici* Leonian was also identified as a causal agent of root rot of lima bean in the United States (Davidson et al., 2002).

The symptoms typically observed are seedling lodging, crown, hypocotyl and root rot and vascular necrosis, resulting in wilting, stunted growth and yellow, dried leaves (Paula Júnior et al., 1995).

The control of these fungi is extremely difficult because they have high survivability in soils and crop residues, primarily through the production of resistance structures, and show widespread occurrence in Brazilian fields. Dissemination occurs via contaminated seeds, rain and irrigation water, humans, animals and farm equipment (Paula Júnior et al., 1995).

Crop rotation with grasses should be implemented for disease control to reduce the inoculum in contaminated areas. Watering should be controlled to prevent excessive soil moisture. The use of healthy treated seeds is also recommended (Paula Júnior et al., 1995).

Bacterial Diseases

Reports of bacterial diseases in lima bean are rare in Brazil. The main bacterial diseases reported worldwide will be described below.

Common Bacterial Blight

Common bacterial blight is caused by the bacterium *Xanthomonas axonopodis* pv. *phaseoli*. This disease may become a serious issue for several legumes, including the common bean and lima bean, under high humidity conditions, although there are no reports of natural occurrence of this pathogen in lima bean in Brazil (Vieira, 1992). Further, there are no reports of lima bean production losses, although common bean blight is considered the most important bacterial disease. In common bean, the typical disease symptoms are visible throughout the entire plant shoot. Small spots with a soggy, translucent appearance are initially observed on the leaves. They subsequently become brown and enlarged, and the infected tissues become parched and brittle, usually with a chlorotic halo, which is typical of many bacterial diseases. The lesions in stems and pods are often depressed and soggy; they gradually increase in size and may become reddish. The lesions may exude yellowish bacterial pus. The symptoms are not always noticeable in seeds, and discoloration may occur only at the hilum (Paula Júnior et al., 1995).

The most effective bacterial blight control results may be achieved with a combination of measures, including the use of pathogen-free seeds, resistant varieties (when possible) and crop rotation. Spraying of chemicals has not shown satisfactory results under field conditions (Torres et al., 2009).

Halo Bacterial Blight of Common Bean

Halo bacterial blight is a typical disease of common bean plants that is caused by the bacterium *Pseudomonas savastanoi* pv. *phaseolicola*, which can also infect cowpea, lima bean, pea and other legumes. Halo bacterial blight was reported in Brazil in 1998 (Yorinori et al., 1998) and may decrease common bean yield by up to 40% (Trindade et al., 2003).

Up to seven bacterial strains that have naturally infected several legumes, including lima bean, common bean and *Phaseolus coccineus*, have been found in South Africa (Fourie, 1998). The symptoms are very similar to those caused by common blight. Following infection, small dark green spots with a soggy appearance appear on the lower leaf surface. The spots progress and coalesce, forming large lesions that eventually become necrotic. Green-yellow halos form around the lesions. The symptoms in stems are also spots with a soggy appearance, sometimes with depressed lesions, that gradually expand and become brown. Lesions that rupture superficially have bacterial exudate and are more common near the first node. The symptoms in pods consist of dark green, soggy lesions that may also coalesce and become brown. The vascular system of pods usually becomes infected, reaching the surrounding tissues and resulting in infection of seeds through the funiculus. Seeds become wrinkled and discoloured when infected at the development stage (Trindade et al.,

2003). The use of resistant varieties and healthy seeds are the control methods most used in common bean. The use of copper-based products enables the reduction of disease onset and significantly increases production (Trindade et al., 2003). There are no studies on the control of this disease in lima bean.

Brown Spot Disease

Brown spot disease is caused by the bacterium *Pseudomonas syringae* pv. *syringae*; it was first reported in the United States (Thaung; Walker, 1957). The pathogen may be introduced into new areas through contaminated seeds. Symptoms in common bean leaves begin as small, soggy spots that develop into brown, necrotic spots 3 to 8 mm in diameter, often with a diffuse, yellow outline; the lesions may coalesce. Depressed brown spots may form in pods, causing deformation if present at the start of pod development (Thaung; Walker, 1957).

Bacterial Wilt Disease

Curtobacterium flaccumfaciens pv. *flaccumfaciens* (Cff) is a systemic pathogen that is the causal agent of the bacterial wilt of *Phaseolus* spp. (Schuster; Sayre, 1967). The typical symptoms of the disease in common bean are primarily wilt, vascular system browning and plant death. Infected plants may not show symptoms under mild temperature conditions (EPPO, 2011). No effective chemical disease control methods have been reported. The best disease control method is the use of certified seeds because Cff is transmitted through seeds. Economically, the most important Cff host is the common bean, although the bacterium also infects lima bean and other legumes (OEPP, 2011). In Iran, Cff has recently reported causing bacterial wilt in *P. lunatus* (Osdaghi, 2014).

Other Diseases

Wildfire disease, caused by *Pseudomonas syringae* pv. *tabaci*, is capable of inducing yellow halo symptoms in inoculated lima bean plants and other legumes (Ribeiro et al., 1979).

Diseases Caused by Nematodes

Root-Knot

There are few reports of natural infection of lima bean plants by nematodes in Brazil. Only three *Meloidogyne* species have been reported: *M. incognita*, *M. javanica* and *M. enterolobii* Yang & Eisenback (sin. *Meloidogyne mayaguensis* Rammah & Hirschmann, 1988; Freire; Ponte, 1976; Manso et al., 1994). The main symptoms are reduced plant size, the formation of numerous root knots and reduced root systems followed by root necrosis. The affected roots become susceptible to infection by secondary pathogens. The root knots are more numerous and larger when parasitised by *M. enterolobii* compared to plants infested by *M. incognita* (Bitencourt; Silva, 2010).

Root-knot symptoms were reported in 10 accessions tested in research studies on accessions from the Lima Bean Active Germplasm Bank of the UFPI. The diagnosis performed using α-esterase electrophoresis of females identified *M. incognita* as the causal agent of the disease (personal communication).

The control of root-knot disease in lima bean is not easy. Producers must coexist with the nematode in areas where it has already been identified because its eradication is practically impossible. However, some measure help to minimise damage, especially crop rotation with non-host plants and the use of resistant varieties when available. Studies identifying sources of lima bean resistant to root-knot nematodes are incipient. Bitencourt; Silva (2010) identified the genotypes UFPI-464, Raio de Sol and UFPI-463 as highly resistant to *M. incognita.* Varieties resistant to *M. incognita* and *M. javanica* have already been reported in California, United States (ANR, 2014).

Final Considerations

In Brazil, lima bean has merited little attention from research institutions, resulting in limited knowledge of various crop aspects, including those related to diseases.

The scarcity of disease reports most likely results from the reduced economic exploitation of the crop. Although lima bean has a lower incidence of diseases than common bean (Vieira, 1992), some diseases, including anthracnose, viral diseases, web blight and root-knot disease, may become an issue with lima bean crop expansion. Common bean diseases, including white mould, angular leaf spot, web blight and Fusarium wilt, began to occur with greater intensity following cultivation of the crop in large irrigated areas (Vale et al., 1997). Thus, to expand lima bean cultivation, studies are needed for effective knowledge of the diseases with the highest incidence in each region, the contribution of those diseases to the loss of crop yield and the etiologic agents of diseases.

REFERENCES

Alves, A.U., Oliveira, A.P., Alves, A.U., Dornelas, C.S.M., Alves, E.U., Cardoso, E.A., Oliveira, A.N.P., Cruz, I.S., (2008). Lima beans production and economic revenue as function of organic and mineral fertilization. *Horticultura Brasileira* 26, 251-254.

Andrade, G.P., Silva, J.L., Pio-Ribeiro, G., Alves-Junior, M., Rodrigues, A.A.C., (2001). Infecção natural de fava com CABMV em Itambé e Terezinha no Estado de Pernambuco [Natural infection of fava bean with cowpea aphid-borne mosaic virus (CABMV) in Itambé and Terezinha in the State of Pernambuco.]. *Fitopatologia Brasileira* 26, 511-512.

ANR, (2014). Publication 8505. Lima bean production in California. Available at: <http://anrcatalog.ucanr.edu>. Accessed on January 06, 2015.

Aragão, F.J.L., Nogueira, E.O.P.E.L., Tinoco, M.L.P., Faria, J.C., (2013). Molecular characterization of the first commercial transgenic common bean immune to the Bean golden mosaic virus. *Journal of Biotechnology* 166, 42-50.

Assunção, I.P., Nascimento, L.D., Ferreira, M.F., Oliveira, F.J., Michereff, S.J., Lima, G.S.A., (2011). Reaction of faba bean genotypes to *Rhizoctonia solani* and resistance stability. *Horticultura Brasileira* 29, 492-497.

Bezerra, J.L., Luz, E.D.M., Santos, A.F., (2001). Doenças por *Phytophthora* spp. em outros hospedeiros [Diseases caused by *Phytophthora* spp. in other hosts]. In: Bezerra, J.L. (Editors). Doenças causadas por *Phytophthora* no Brasil [Diseases caused by *Phytophthora* in Brazil]. Campinas (SP): Editora Rural.

Bianchini, A., Maringoni, A.C., Carneiro, S.M.T.P.G., (2005). Doenças do feijoeiro [Bean diseases]. In: Kimati, H., Amorim, L., Resende, J.A.M., Bergamim Filho, A., Camargo, L.E.A. (Editors). Manual de Fitopatologia.

Doenças das Plantas Cultivadas [Phytopathology Handbook. Diseases of Cultivated Plants], v.2, 4th ed. São Paulo: Agronômica Ceres.

Bitencourt, N., Silva, G.S., (2010). Reação de genótipos de fava a *Meloidogyne incognita* e *M. enterolobii* [Reaction of lima bean to *Meloidogyne incognita* and *M. enterolobii*]. *Nematologia Brasileira* 34, 184-186.

Bonde, M.R., Nester, S.E., Berner, D.K., Frederick, R.D., Moore, W.F., Little, S., (2008). Comparative susceptibilities of legume species to infection by *Phakopsora pachyrhizi*. *Plant Disease* 92, 30-36.

Buchwaldt, L., Anderson, K.L., Morrall, R.A.A., Gossen, B.D., Bernier, C.C., (2004). Identification of lentil germplasm resistant to *Colletotrichum truncatum* and characterization of two pathogen races. *Phytopathology* 94, 236-243.

Cavalcante, G.R.S., Carvalho, E.M.S., Gomes, R.L.F., Santos, A.R.B., Santos, C.M.P.M., (2012). Reação de subamostras de feijão-fava à antracnose [Lima bean subsample reaction to anthracnose]. *Summa Phytopathologica* 38, 329-333, 2012.

Carvalho, E.M.S. Antracnose em feijão-fava: caracterização do agente causal e reação de genótipos a *Colletotrichum truncatum*. 2009. 53p. Tese (Doutorado em Agronomia/Produção Vegetal) - Universidade Estadual Paulista, Faculdade de Ciências Agrárias e Veterinárias, Jaboticabal, 2009 [Anthracnose in lima bean: characterization of the causal agent and reaction of genotypes to *Colletotrichum truncatum*. 2009. 53p. Thesis (PhD in Agronomy/Plant Production) - São Paulo State University, School of Agricultural and Veterinary Sciences, Jaboticabal, 2009].

Chel-Guerrero, L., Gallegos-Tintoré, S., Martínez-Avala, A., Castellanos-Ruelas, A., Betancur-Ancona, D., (2011). Functional properties of proteins from Lima bean (*Phaseolus lunatus* L.) seeds. *Food Sci. Techonol. Int.* 17, 119-126.

Coco, D., Calil, I.P., Brustolini, O.J.B., Santos, A.A., Inoue-Nagata, A.K., Fontes, E.P.B., (2013). Soybean chlorotic spot virus, a novel begomovirus infecting soybean in Brazil. *Archives of Virology* 158, 457-462.

Costa, A.F. Efeito de infecções simples e duplas de quatro vírus do feijoeiro (*Phaseolus vulgaris* L.) na produção e em algumas características da planta. 1983. 109p. Dissertação de Mestrado) – Universidade de Brasília, Brasília, 1983 [Effect of single and double infections of four viruses in common bean plants (*Phaseolus vulgaris* L.) on production and some plant characteristics. 1983. 109p. (MSc. Thesis) – University of Brasília, Brasília, 1983].

Costa, A.F., (1995). Doenças causadas por vírus em hortaliças leguminosas. Doenças de hortaliças [Viral diseases in legumes. Vegetable diseases]. *Informe Agropecuário* 17, 75-85.

Costa, C.L., Matos, F.S.A., Santos, O.R., Kitajima, E.W., Meyer, M.C., (1991a). Um comovírus isolado de planta obtida de semente de *Phaseolus lunatus*, colhida em Jataí, GO [A comovírus isolated from *Phaseolus lunatus* plants grown from seeds, harvested in Jataí, Goiás (GO)]. *Fitopatologia Brasileira* 16, 25.

Costa, C.L., Matos, F.S.A., Santos, O.R., Kitajima, E.W., (1991b). Um potyvírus isolado de feijão de lima (*Phaseolus lunatus*) [Potyvirus isolated from lima bean (*Phaseolus lunatus*)]. *Fitopatologia Brasileira* 16, 27.

Costa, A.F., (1986). Antracnose da fava (*Phaseolus lunatus* L.) no nordeste causada por *Colletotrichum dematium f. truncata* [Lima bean anthracnose (*Phaseolus lunatus* L.) in the Brazilian Northeast caused by *Colletotrichum dematium f. truncata*]. *Fitopatologia Brasileira* 11, 315.

Davey, J.F., Gregory, N.F., Mulrooney, R.P., Evans, T.A., Carroll, R.B., (2008). First report of mefenoxam-resistant isolates of *Phytophthora capsici* from lima bean pods in the mid-Atlantic region. *Plant Disease* 92, 656.

Davidson, C.R., Carrol, R.B., Evans, T.A., Mulrooney, R.P., Kim, S.H., (2002). First Report of *Phytophthora capsici* infecting lima bean (*Phaseolus lunatus*) in the mid-Atlantic region. *Plant Disease* 86, 1049.

Davidson, C.R., Evans, T.A., Mulrooney, R.P., Gregory, N.F., Carroll, R.B., O'Neill, N.R., (2008). Lima bean downy mildew epiphytotics caused by new physiological races of Phytophthora phaseoli. *Plant Disease* 92, 670-674.

Embrapa Soja, (2008). Tecnologias de Produção de Soja – Região Central do Brasil - 2009 e 2010 [Soy Production Technologies - Central Region of Brazil - 2009 and 2010]. Sistema de Produção n° 13, Brasília.

EPPO, (2011). *Curtobacterium flaccumfaciens pv. flaccumfaciens*. OEPP/EPPO Bulletin 41, 320-328.

Evans, T.A., Davidson, C.R., Dominiak, J.D., Mulrooney, R.P., Carrol, R.B., (2002). Two new races of *Phytophthora phaseoli* from lima bean in Delaware. *Plant Disease* 86, 813.

Evans, T.A., Mulrooney, R.P., Gregory, N.F., Kee, E., (2007). Lima bean downy mildew: impact, etiology, and management strategies for Delaware and the mid-Atlantic region, U. S. *Plant Disease* 91, 128-135.

Fernandes, F.R., Cruz, A.R.R., Faria, J.C., Zerbini, F.M., Aragão, F.J.L., (2009). Three distinct begomoviruses associated with soybean in central Brazil. *Archives of Virology* 154, 1567–1570.

Figueiredo, M.B., (2001). Doenças fúngicas emergentes em grandes cultivares (Palestra) [Emerging fungal diseases in large cultivars (Lecture)]. *Biológico* 63, 29-32.

Fourie, D., (1998). Characterization of halo blight races on dry beans in South Africa. *Plant Disease* 82, 307-310

Freire, F.C.O., Ponte, J.J., (1976). Nematóides das galhas, *Meloidogyne* spp. associados ao parasitismo de plantas no Estado da Bahia [Gall nematodes, *Meloidogyne* spp. associated with plant parasitism in the State of Bahia]. *Boletim Cearense de Agronomia* 17, 47-55.

IBGE. Banco de dados agregados: pesquisa: produção agrícola municipal, 2013 [Aggregated database: research: municipal agricultural production, 2013]. Available at <http://www.sidra.ibge.gov.br/bda/>. Accessed on January 12, 2015.

Lemos, L.B., Fornasieri Filho, D., Silva, T.R.B., Soratto, R.P., (2003). Suscetibilidade de genótipos de feijão ao vírus-do-mosaico-dourado [Susceptibility of common bean genotypes to gold mosaic virus]. *Pesquisa Agropecuária Brasileira* 38, 575-581.

Lima, J.A.A., Santos, C.D.G., (1985). Infecção natural de *Phaseolus lunatus* por uma estirpe de "cucumber mosaic virus", no Ceará [Natural infection of *Phaseolus lunatus* by a strain of cucumber mosaic virus, in Ceará]. *Fitopatologia Brasileira* 10, 304.

Lima, J.S., Assunção, I.P., Teodoro, I., Lima, G.S.A., Michereff, S.J., (2011). Influência do sistema de irrigação na incidência e nas perdas ocasionadas pelo mosaico dourado em feijoeiro [Influence of irrigation systems on incidence and losses caused by golden mosaic in common bean]. *Tropical Plant Pathology* 36, 50-53.

Lynch, T.N., Marois, J.J., Wrigt, D.L., Harmon, P.F., Harmon, C.L., Miles, M.R., Hartman, G.L., (2006). First report of soybean rust caused by *Phakopsora pachyrhizi* on *Phaseolus* spp. in the United States. *Plant Disease* 90, 970.

Manso, E.C., Tenente, R.C.V., Ferraz, L.C.C.B., Oliveira, R.S., Mesquita, R., (1994). Catálogo de Nematóides Fitoparasitos Encontrados Associados a Diferentes Tipos de Plantas no Brasil [Catalogue of Phytoparasitic Nematodes Found in Association with Different Plant Types in Brazil]. Embrapa – Centro Nacional de Pesquisa de Recursos Genéticos e Biotecnologia, Brasília, 488 p.

Melgarejo, T.A., Lehtonen, M.T., Fribourg, C.E., Rännäli, M., Valkonen, J.P.T., (2007). Strains of BCMV and BCMNV characterized from lima bean plants affected by deforming mosaic disease in Peru. *Archives of Virology* 152, 1941-1949.

Nechet, K.L., Halfeld-Vieira, B.A., (2006). Severidade da mela (*Rhizoctonia solani*) em genótipos de feijão-caupi em Roraima [Severity of web blight in cowpea cultivars (Rhizoctonia solani) in Roraima, Brazil]. *Fitopatologia Brasileira* 31, 374.

Nechet, K.L., Halfeld-Vieira, B.A., Gianluppi, V., Meyer, M.C., (2004). Avaliação de genótipos de soja em relação a antracnose (*Colletotrichum truncatum*) e mela (*Thanatephorus cucumeris*) nas condições de Roraima [Evaluation of soybean genotypes for anthracnose (*Colletotrichum truncatum*) and Web blight (*Thanatephorus cucumeris*) under the conditions in Roraima]. *Fitopatologia Brasileira* 29, 35.

Osdaghi, E. (2014). Bacterial wilt of lima bean (*Phaseolus lunatus*) caused by *Curtobacterium flaccumfaciens* pv. *flaccumfaciens*, a new disease in Iran. *Journal of Plant Pathology* 96, S4.113-S4.131.

Paula Júnior, T.J., Silva, M.B., Vieira, R.F., (1995). Doenças causadas por fungos em hortaliças leguminosas [Fungal diseases in legume vegetables]. *Informe Agropecuário* 17, 63-85.

Pereira, D.R. Fatores que influenciam o desenvolvimento da antracnose do feijão-fava [Factors influencing the development of anthracnose in lima bean]. 2014. 51p. Dissertação (Mestrado em Agronomia/Produção Vegetal) - Universidade Federal do Piauí [Dissertation (MSc. in Agronomy/Plant Production) - Federal University of Piauí], Teresina, 2014.

Ramos-Sobrinho, R., Xavier, C.A.D., Pereira, H.M.B., Lima, G.S.A., Assunção, I.P., Mizubuti, E.S.G., Duffy, S., Zerbini, F.M., (2014). Contrasting genetic structure between two begomoviruses infecting the same leguminous hosts. *Journal of General Virology* 95, 2540-2552.

Ribeiro, R.L.D., Hagedorn, D.J., Durbin, R.D., Uchytil, T.F., (1979). Characterization of the bacterium inciting bean wildfire in Brazil. *Phytopathology* 69, 208-212.

Santos, D., Corlett, F.M.F., Mendes, J.E.M.F., Wanderley Júnior, J.S.A., (2002). Produtividade e morfologia de vagens e sementes de variedades de fava no Estado da Paraíba [Productivity and morphology of pods and seeds of lima bean varieties in Paraíba State, Brazil]. *Pesquisa Agropecuária Brasileira* 37, 1407-1412.

Schuster, M.L., Sayre, R.M., (1967). A coryneform bacterium induces purple-colored seed and leaf hypertrophy of *Phaseolus vulgaris* and other leguminosae. *Phytopathology* 57, 1064-1066.

Sengooba, T.N., Spence, N.J., Walkey, D.G.A., Allen, D.J., Femi Lana, A., (1997). The occurrence of bean common mosaic necrosis virus in wild and forage legume in Uganda. *Plant Pathology* 46, 95-103.

Shahid, M.S, Ikegami, M., Natsuaki, K.T., (2012). First report of *Mungbean yellow mosaic India virus* on Lima bean affected by yellow mosaic disease in Nepal. Australasian *Plant Disease Notes* 7, 85–89.

Silva, J.A., Oliveira, M.G., Souza, L.T., Assunção, I.P., Lima, G,A,, Michereff, S,J., (2014). Reação de genótipos de feijão-fava a *Sclerotium rolfsii* [Reaction of fava bean genotypes to Sclerotium rolfsii]. *Horticultura Brasileira* 32, 98-101.

Silva, K.J.D., Lopes, A.C.A., Gomes, R.L.F., Sousa, F.M., Rocha, M.M., (2010). Recursos genéticos [Genetic resources]. In: LOPES ACA; GOMES RLF; ARAÚJO ASF (eds). A cultura do feijão-fava no Meio-Norte do Brasil [The fava bean crop in the Brazilian Mid-North]. Teresina: Editora da UFPI. 272p.

Slaminko, T.L., Miles, M.R., Frederick, R.D., Bonde, M.R., (2008). New legume hosts of *Phakopsora pachyrhizi* based on greenhouse evaluations. *Plant Disease* 92, 767-771.

Thaung, M.M., Walker, J.C., (1957). Studies on bacterial blight of lima bean. *Phytopathology* 47, 413-417.

Torres, J.P., Silva Júnior, T.A.F., Maringoni, A.C., (2009). Detecção de *Xanthomonas axonopodis* pv. *phaseoli* em sementes de feijoeiro provenientes do estado do Paraná, Brasil [Detection of *Xanthomonas axonopodis* pv. *phaseoli* in common bean seeds from the state of Paraná (Brazil)]. *Summa Phytopathologica* 35, 136-139.

Trindade, L.C., Coelho, M.V.S., Marques, A.S.A., (2003). Crestamento bacteriano aureolado do feijoeiro: praga quarenternária para o Brasil com alto risco de disseminação no país [Halo bacterial blight: quarantine pest in Brazil at high risk of dissemination in the country]. Comunicado Técnico. (EMBRAPA), n°100, DEZ, p.1-10.

Vale, F.X.R., Costa, H., Zambolim, L., (1997). Feijão comum (*Phaseolus vulgaris* L.) Controle de doenças: doenças de parte aérea causada por fungos [Common bean (*Phaseolus vulgaris* L.) Disease control: shoot fungal diseases]. In: Vale, F.X.R.; Zambolim, L (Editors). Controle de doenças de plantas [Plant disease control]. Viçosa, MG: UFV,

Departamento de Fitopatologia; Brasília, DF: Ministério da Agricultura e do Abastecimento, v.1, p.335-374.

Vieira, R.F., (1992). A cultura do feijão-fava [The fava bean crop]. *Informe Agropecuário* 16, 30-37.

Vieira, R.F., Vieira, C., Andrade, G.A., (1992). Comparações agronômicas de feijões dos gêneros *Vigna* e *Phaseolus* com o feijão-comum (*Phaseolus vulgares* L.) [Agronomic comparison of *Phaseolus* and *Vigna* genera beans with the common bean (*Phaseolus vulgares* L.)]. *Pesquisa Agropecuária Brasileira* 27, 841-850.

Yorinori, M.A., Funada, C.K., Rodrigues, R., Bianchini, A., Leite, Jr., R.P., Ueno, B. (1998). Ocorrência do crestamento bacteriano aureolado do feijoeiro causado por *Pseudomonas syringae* pv. *phaseolicola* no Estado do Paraná [Occurrence of halo bacterial blight in bean plants caused by *Pseudomonas syringae* pv. *phaseolicola* in the State of Paraná]. *Fitopatologia Brasileira* 23, 218.

In: Phaseolus Lunatus
Editors: A. S. Ferreira de Araujo et al.

ISBN: 978-1-63482-494-1

Chapter 9

Aspects of Crop and Socioeconomic Management of the Lima Bean within Northeastern Brazil

Regina Lucia Ferreira Gomes[1,*], Cristiane Araújo Soares[2], Francisco Carlos Gândara[1] and Marília Lobo Burle[3]

[1]Universidade Federal do Piauí, Campus Ministro Petrônio Portela, s/n, Bairro Ininga, CEP 64.049-550, Teresina, Piauí, Brazil
[2]Universidade Estadual Paulista " Julio de Mesquita Filho", Campus de Botucatu, São Paulo, Brazil.
[3]Empresa Brasileira de Pesquisa Agropecuária, Embrapa-Cenargen, Brasília, DF, Brazil

Abstract

Socioeconomic studies comprise an effective tool that captures the variables related to the production process. From this perspective, it is essential to understand the socioeconomic and environmental diagnoses of lima bean production in northeastern Brazil, considering that this legume is consumed in many different forms and is very important for human and animal feeding. Regarding the lima bean production system, the technology used is simple and rudimentary; there is no type of soil

* Corresponding author: Email: rlfgomes@ufpi.edu.br.

conservation management, and agricultural chemicals are not used for pest and disease control in crops, which is instead controlled naturally. Although farmers use this legume for consumption, a significant portion of them also sell the crop mainly within the municipality where they are grown. Regarding the aspects of the income composition of the farmers, most receive public policy contributions, and 90% do not reside within the production unit because most are not the landowners. There is a higher proportion of men working with lima bean crops, and most of them are married and live in brick houses, in which often there is no basic sanitation. Thus, creating development strategies more adequate for the lima bean farming conditions of northeastern Brazil would contribute to the improvement of the quality of life of the rural communities.

Keywords: *Phaseolus lunatus*, grain production, phytotechnical practices, infrastructure

INTRODUCTION

The lima bean (*Phaseolus lunatus* L.) is one of the main legumes farmed in many tropical countries; it better adapts to hot and humid climates and should be preferably farmed during the low rainfall period. Its use is lower than that of the common bean; however, it seems to have wider adaptability (Lymman, 1983), considering it is more drought tolerant and tolerant to excess moisture and heat (Vieira, 1992).

The lima bean is very important in human and animal food and is used as green manure and as a cover crop, to protect the soil from erosion caused by heavy rainfall. The forms of lima bean consumption vary, where dry or green beans, green pods, and leaves are used as human food (Filgueira, 2000). The United States is one of the largest lima bean producers worldwide, where the legume is eaten in its green state; it is very well liked in its preserved form (canned or frozen and packaged beans), exceeding the consumption of its dry bean form (Vieira, 1992). In northeastern Brazil, dry bean consumption prevails, although green lima beans are sold at farmers markets and public markets, reaching higher prices than the dry beans, thus ensuring the survival of farmers that use it as a source of food and income (IBGE, 2014).

For the human population, this vegetable has the potential to provide plant protein, functioning as an alternative source of food in several municipalities of northeastern Brazil, together with cowpea, corn, and cassava (Vieira, 1992), considering the greater difficulty of accessing animal protein. The dry beans comprise 62.9% carbohydrates, 25% protein, 6.1% fibre, 3.9% ash, and 2.0% ether extract (Bressani and Elias, 1980). Due to its protein content and characteristic taste, it is used in dishes worldwide, in many different cuisines, being given several names, depending on the region farmed or the way it is used for consumption (Santos, 2008).

Among the legumes that produce hydrogen cyanide (HCN), only lima beans contain this chemical in high quantity; this characteristic is responsible for its bitter taste and is lacking in other bean species (Vieira, 1992). The toxicity of some genotypes is only characterised by their bitter taste; to eliminate it, lima beans must be cooked three to five times, completely replacing the water used (Azevedo et al., 2003). Even with these precautions, lima beans with high HCN content can be toxic if consumed regularly (Vieira, 2007). Toxic levels ranging from 15-25 ppm to 115-150 ppm were found in seven lima bean varieties sold at CEASA-PI (Azevedo et al., 2003).

The lima bean crop may also function as forage plants, producing large quantities of moderate quality forage, which is adequately accepted by cattle. The forage value of the green plants is limited due to the presence of linamarin, which releases HCN. Although this glucoside forms during the last maturation steps, it is concentrated in younger branches and in seeds, particularly in the more stained seeds. Dark seeds contain fewer toxins than stained ones; however, they have a bitter taste, whereas light seeds are linamarin free (Beyra and Artiles, 2004). Similar to other tropical legumes, lima bean seeds contain some anti-nutritional factors, which limit their use in animal feed (Ologhobo and Fetuga, 1983; Santos, 2008).

Lima beans have received little attention by some research and extension agencies, resulting in limited knowledge of its agronomic characteristics. In northeastern Brazil, it is very economically and socially important because it causes robustness, having its harvest prolonged and performed during the dry season. Lima bean farming in this region is rustic, intercropped with corn, cassava, or castor, using plants of these crops as support (Azevedo et al., 2003).

The main reasons for the relatively limited lima bean farming are believed to be the tradition of common bean consumption, taste of lima beans, longer cooking time, and the lack of varieties adapted to the farming conditions of each region (Lymman, 1983). Low productivity can be attributed to a portion of the production coming from small-scale farmers who use intercropped systems without adopting technologies aimed at increased productivity, such as using improved varieties (Vieira, 1992).

On *P. lunatus* L. germplasm sampling expeditions, a study was performed by applying the following: 50 questionnaires in 12 municipalities of Piauí state (Água Branca, Barra d'Alcântara, Barro Duro, Hugo Napoleão, José de Freitas, Novo Oriente do Piauí, Olho d'Água do Piauí, Palmeirais, São Pedro do Piauí, Tanque do Piauí, Teresina, and Várzea Grande) from August to November 2008; 10 questionnaires in five municipalities of Ceará (Crateús, Iguatú, São Benedito, Tianguá, Viçosa) from October to November 2011; seven questionnaires in two municipalities of Bahia (Heliópolis and Euclides da Cunha) from February to March 2012; and 34 questionnaires in 15 municipalities of Paraíba (Alagoa Nova, Arara, Borborema, Casserengue, Esperança, Ingá, Matinhas, Mogeiro, Montadas, Puxinanã, Queimadas, Remígio, São Tomé de Lagoa da Roça, Serraria, and Solânea) from September 2009 and September 2014. This study included the highest lima bean producers of the respective states (Soares et al., 2009; IBGE, 2014). Productive and socioeconomic aspects regarding the lima bean crops in northeastern Brazil were addressed in these questionnaires.

Crop Management

Lima beans are farmed in many tropical countries from the sea level until an altitude of 2,400 m (Duke, 1983). The optimal temperature for seed germination varies between 24°C and 30°C; a low temperature (below 13°C) slows growth, and a high nocturnal temperature (above 21°C) accelerates plant maturation and limits the number and size of the seeds. There is high variability in the lima bean response to photoperiod, where some cultivars are neutral; others are short-day, flowering within the critical photoperiod from 11 to 12.5 hours (Baudoin, 1988).

Lima beans can be grown throughout the year in low and hot areas (Filgueira, 2000). Dryland (rainfed) agriculture is predominately used in northeastern Brazil, where sowing is performed at the onset of the rainy period. The lima bean adapts well to these conditions, tolerating the poor rainfall distribution that is common within the region.

The cultivated forms of lima beans are annual or perennial, climbers or shrubs, with little variation regarding the type of plant and life cycle. Climbing types can reach 4.0 m in height (Duke, 1983). Most of the lima bean cultivars used in Brazil demonstrate a late life cycle, lasting approximately 6 months, an indeterminate growth habit, climbing ability, and uneven maturation, where harvest is performed in several passes (Vieira, 1978).

Within northeastern Brazil, because the lima bean is a climbing plant, its farming is restricted to the space occupied by remaining trees—usually the babassu palm—after clearing has been performed, using them as support for its vegetative development and making harvesting difficult and dangerous. It is also farmed by intercropped planting. This was the only planting system adopted, confirming intercropping as one of the causes for the low lima bean yield in Brazil. In intercropping, the lima bean participates as a marginal crop, planted close to the trees or at the margins of the clearings. In this case, fences or other crops such as castor, cassava, and corn are used as support. This form of exploitation demonstrates the rudimentary way of the production system used and reveals the little importance attributed to the lima bean crop by a portion of the research and rural extension agencies; consequently, limited knowledge of their agronomic characteristics is known (Santos et al., 2002).

Usually, the following main agricultural production systems are observed for the lima bean crop: intercropping with rice, corn, cassava and lima bean, or rice, corn and lima bean, farmed using slash and burn, using rudimentary technology, with seeds obtained from the same field and irregular variation in spacing and intercropped arrangements.

In Gorutuba, MG, Brazil, a study on the performance of 15 lima bean cultivars intercropped with corn, during the 1984/85 and 1985/86 crop years, with rows spaced 1.0 m apart, holes spaced 0.4 m apart, and a density of two plants per hole, found that lima beans exhibited a mean yield of 1,612 and 2,086 kg of dry beans/ha in the first and second trials, respectively, and corn exhibited yield a little higher than 2,500kg/ha (Vieira, 1992).

The production process performed in the rural zone of northeastern Brazil usually comprises simple practices, without any type of technological innovation, using techniques handed down from generation to generation, such as a heritage from ancestors. Among the farmers interviewed, few use techniques of soil preparation and organic fertilisation with cattle manure. In addition, they do not use soil conservation practices, and the soil is abandoned after harvest and left to fallow for several years.

In the management of the productive sustainability of soil, the means available to improve the chemical conditions are based on applying fertilisers and correctives; however, adopting these practices results in expressive increases in production costs, making it inaccessible for many family farmers. Thus, it is necessary to adopt alternative practices that are accessible to the farmers and that can help maintain the productive capacity of soil intended for agricultural activity (Ramos, 2001).

The vast majority of the farmers that grow lima beans practice burning on the properties, a technique that is one of the steps to prepare the area for planting in a process referred to as "slash and burn". In the latter process, the agricultural production model is based on the annual use of at most 2 ha per family, using the processes of clearing the vegetation and burning, which occur rudimentarily and result in low productivity indices. In Maranhão State, family agriculture production systems comprise an intercropping of three or more crops in uncleaned clearings. It is a shifting cultivation agriculture system, in which, after a farming cycle, the area is maintained at rest for 6-10 years for the vegetation to regenerate (Oliveira Junior et al., 2007).

In some regions, low productivity levels have been found where the lack of a research program with mineral nutrition is one of the main causes. The quantities of phosphorus required by the vegetable are usually low, particularly compared with nitrogen and potassium. However, supplying it in adequate doses favours root system development (increasing the absorption of water and nutricnts), increases the vigour of the plants from direct seeding, favours flowering and fruiting, and increases the quality and yield of the products harvested (Filgueira, 2000). However, to date, little is known regarding the quantities to be used, allowing for the acquisition of a satisfactory lima bean yield.

In the lima bean crop in phosphorus-poor soils, using doses above 1,500 kg/ha of a single superphosphate is unnecessary. The "Orelha de Vó" cultivar obtained a maximum production of green and dry beans estimated by derivations of 5.2 and 2.7 $t.ha^{-1}$, respectively, when using 302 and 309 kg ha^{-1} of P_2O_5, respectively (Oliveira et al., 2002).

The low lima bean productivity can also be attributed to the lack of studies on organic fertilisation that caused a low yield and economic return, making the lima bean practically a subsistence crop, where only surplus production was sold in open farmers markets (Frazão et al., 2004).

In the agricultural production model developed by small-scale farmers using lima bean as a source of food and income, few use agricultural chemicals for pest control and diseases in crops, where control is thus performed naturally, via homeostatic mechanisms (self-regulation) (Gândara, 1998). One very worrying fact is that, of the 40% that use these products, only a quarter of them know the product's name, and many of them do not even know how to define its function.

In studying the form of occurrence at the field level of the propagation material, use of creole cultivars was revealed, showing a lack of improved cultivars in the clearings. The beans exhibit varieties of colours and sizes, indicating the great genetic variability available. Thus, as planting several materials within the same area is common, it is most likely that many samples come from a natural hybridisation process because the natural hybridisation rate can range from 4% to 100% (Rachie et al., 1980).

As it is a rainfed crop, planting lima beans in this region coincides with the onset of the rainy period, occurring in December and January, different from other regions, when sowing occurs from September to March (Filgueira, 2000).

Regarding the area planted and soil type, most lima beans are farmed in clayey soil (41%) and in highlands (65%), where two to four beans are distributed per hole without thinning. Lima beans grow best in sandy-clayey, fertile, well-drained soil, with a pH between 5.6 and 6.8, distributing three to four beans per hole, leaving two plants (Filgueira, 2000). Regarding the flavour, 96% of the samples do not exhibit a bitter taste or it is weak, indicating a process of natural selection performed by farmers themselves, and only 4% exhibited a medium bitter taste.

Among the main phytosanitary problems observed by farmers, in farmed areas, the occurrence of anthracnose (*Colletotrichum truncatum*) was reported. There is an association between anthracnose and the stergasta caterpillar, which consumes the seed while it is still in the pod, *Chalcodermus bimaculatus* Fiedler and weevils (*Zabrotes subfasciatus*).

All of the genotypes collected in Piauí and Bahia demonstrate an indeterminate growth habit, climbing ability, a late life cycle, and uneven maturation. They are planted using the slash and burn technique and are planted together with other crops. Among the creole cultivars collected in Paraíba, 77.4% exhibit an indeterminate growth habit as well as climbing ability, and 22.6% have a determinate habit and shrubby behaviour. In Ceará, 90% of the cultivars collected have an indeterminate habit as well as climbing ability.

Regarding lima bean storage, most of the farmers interviewed report using polyethylene terephthalate bottles, particularly for the portions intended for the next planting. Those intended for consumption and sale, when in higher quantity, are placed into nylon bags and/or drums. Some farmers usually place the seeds in containers with sand to help preserve the seeds, hindering attack of the stored grains by pests, a situation that is one of the largest problems faced in lima bean storage. In Paraíba, there was a report of using ash and black pepper to control these pests.

Family agriculture is responsible for producing a large portion of the food consumed by the Brazilian population; however, there is much loss due to the inadequate conditions in which they are stored, making many tons of product unfit for consumption and sale. In bean storage, much has been done to reduce the losses and adequately conserve the agricultural products because any changes in the quality or availability of the food are immediately apparent to humans, leading to important socioeconomic repercussions; these beans are susceptible to infestation by pests and diseases, and chemical control is the treatment type most widely used, at times leaving toxic residues in the human organism (Silva et al., 2005).

ECONOMIC AND SOCIAL ASPECTS

In Brazil, in 2012, 20,969 tons of dry lima beans were produced in a planted area of 27,034 ha. The states of Ceará, Pernambuco, Paraíba, Piauí, Sergipe, Maranhão, Alagoas, and Rio Grande do Norte, Brazil, are the largest producers, in decreasing order (Table 1). Together, they make northeastern Brazil the largest producing region, with a planted area of 25,872 ha, where 19,979 ha are harvested. Minas Gerais State is the only producer of the southeastern region, with 467 tons, and the southern region only produces 110 tons, in Rio Grande do Sul State (IBGE, 2014).

The socioeconomic transformations that occurred within the northeastern region over the last 50 years were not sufficient to reduce the income inequality and poverty (Mariano and Neder, 2004). The lima bean yields in Brazil and the producing states within a time series between 1990 and 2012 (Table 2) show that there was a very marked decrease in the mean yield in most of the states in 2012, particularly in Piauí and Paraíba. This most likely occurred due to successive years of drought that affected bean productivity throughout the northeastern region. In this region, the lima bean is used as an alternative source of food and income by the population, particularly by family agriculture. Dry bean yield and the price reached with the lima bean may be higher than that found with the common bean. However, the technological resources of production are not available to the farmers or are difficult to obtain, where the largest portion of the production and crop is often intercropped with other crops, in small scattered areas (Vieira and Araújo, 1992).

Table 1. Planted and harvested areas, quantity produced, production, mean yield, and lima bean production value (in beans) in the northeastern states of Brazil, except Bahia, in 2012

States	Planted area (ha)	Harvested area (ha)	Production (t)	Mean yield (kg/ha)	Value 1,000 (R$)
Piauí	2,007	1,891	283	150	1,273
Maranhão	570	564	157	278	1,086
Ceará	7,538	7,538	2,075	275	9,638
Rio Grande Norte	327	10	5	500	15
Paraíba	10,248	5,209	707	136	2,784
Pernambuco	4,490	4087	971	238	2,859
Alagoas	154	123	51	415	247
Sergipe	502	457	206	451	547

Source: IBGE, 2014

Socioeconomic studies comprise a tool that not only effectively captures the variables related to the production process, but also the limitations and/or difficulties in production. Moreover, based on the deep knowledge of the diverse systems in use, by identifying the technological and socioeconomic coefficients, alterations and proposals may be proposed for the alternative models of diversified production based on local conditions and realities (Santos et al., 2002). Socioeconomic and environmental diagnoses cannot be understood as a solution for the problems existing within the region, but they can serve as important tools in creating development strategies more adequate to the conditions of the farmers, contributing to the improvement of the quality of life of the rural communities in the municipalities (Gerhardt et al., 1999).

Regarding the profile of the group interviewed, there is a higher proportion of men who work in lima bean production. Of the 101 farmers, 82.18% are male (57.14% in Bahia, 90% in Ceará, 76.47% in Paraíba, and 88% in Piauí). This can be explained by the women executing other tasks—i.e., domestic activities—in contrast to urban agriculture, which is not performed predominantly by one sex (Pessoa et al., 2006). Regarding the marital status of these farmers, most are married, indicating the needs of family formation and generating income for survival.

Table 2. Time series of the mean yield of the production of the lima bean seasonal crop (kg/ha) in the Brazilian states from 1990 to 2012

Brazil, Region, and State	Year																						
	1990	1991	1992	1993	1994	1995	1996	1997	1998	1999	2000	2001	2002	2003	2004	2005	2006	2007	2008	2009	2010	2011	2012
Brazil	235	268	244	218	207	254	263	284	272	291	382	311	319	371	384	382	405	452	476	457	265	449	240
Northeast	218	260	222	141	178	231	246	277	245	275	376	299	312	367	376	372	396	444	469	414	251	447	224
Maranhão	165	167	139	144	146	145	215	187	187	221	226	327	252	263	280	280	268	265	297	297	298	321	278
Piauí	366	355	251	244	315	321	208	275	154	266	346	184	159	299	330	252	330	184	331	363	236	483	150
Ceará	174	182	181	164	346	235	226	232	297	245	243	85	113	182	112	112	184	244	275	314	125	360	275
Rio Grande do Norte	185	217	149	50	152	187	182	188	117	223	218	206	281	318	352	371	440	427	441	466	269	436	500
Paraíba	280	295	268	115	133	235	225	310	234	264	466	424	411	449	476	477	488	588	568	431	266	445	136
Pernambuco	123	236	145	110	233	234	347	378	383	379	371	309	320	312	331	330	362	380	522	522	420	595	238
Alagoas	205	204	212	259	274	306	299	289	248	244	406	430	521	426	422	374	404	477	459	436	397	429	415
Sergipe	290	260	166	243	182	279	284	273	312	303	279	293	299	406	415	446	440	442	436	459	460	458	451
Bahia	241	348	357	377	379	394	494	200	1,200	1,114	-	-	-	-	-	-	-	-	-	-	-	-	-

Source: IBGE, 2014

The age range of the farmers was between 21 and 75 years, with a mean age of 55 years, and 38% were between 54 and 64 years of age. These data corroborate those of ABMR&A (2005), where the highest percentage of the age range of farmers was above 51 years (31%), and those of Pessoa et al. (2006), for whom the practitioners of urban agriculture were predominantly adult or elderly individuals.

Illiteracy rates in Brazil, which are already very high, become further worrisome in the rural areas (Ramos, 2010). A higher percentage of the farmers interviewed are illiterate, without any level of education. These data may be explained by the lack of conditions to continue studying and the need to start working very early to contribute to household upkeep.

Among the respondents, the mean of the household dependents was between two and three dependents per residence, ranging from one to twelve people. In most of the cases, they lived within the residence. In each family of the small-scale farmers, the mean number of children was four but could reach up to eight. Unemployment and the lack of a fixed income, as well as difficulty to access quality public services, were recalled by some of the farmers interviewed as a limitation for them in having more children (Arco-Verde et al., 2004).

Regarding the infrastructure conditions, among the respondents, 52.63% did not reside within the production unit because most were not landowners. In Piauí, this percentage was 90%, a situation corroborated by the data presented in the 2003/2010 Piauí Agrarian Reform (*Plano de Reforma Agrária do Piauí* 2003/2010), which indicated establishment within smaller areas than the state mean of the settlements, belonging to the following categories: tenant, partner, and occupant. However, in Bahia (85.71%), Ceará (90%), and Paraíba (90.63%), the largest portion of the farmers resided within the production unit.

In analysing the total number of farmers surveyed, 62.38% have their own land; however, in Bahia, Ceará, and Paraíba, the percentages are 100%, 90%, and 85.29%, respectively. In Piauí, less than 35% have their own land, a reality already detected in other studies—i.e., most of the farmers do not own the production unit, where use of the land was possible due to the lease system, representing 62% of the land access conditions. In most cases, this lease is paid with the fixed volume of the given production per unit of the area planted. A very interesting fact is that this rent is usually paid with other crops such as corn and rice, as the quantity of lima bean produced is often not sufficient to pay this lease. Most often, the lease is yearly and a means of obtaining profits by selling the agricultural production (Vieira, 1992). The remaining 6% work and/or live on the land ceded by the owner, who is most often a family member, with a single case where the land was provided by the local government, as if it were a community garden.

The land is not only a productive input but also an asset that may provide other benefits to its owners. By contrast, the rental market depends on the law of supply and demand. Mainly in Latin American countries, some groups have not only a productive opportunity in landholdings that diversifies their investment portfolio but also a great mechanism of protecting against macroeconomic instability determined by sectorial policies (Mendes, 2005).

Most of the respondents live in brick houses, corresponding to 100% in Bahia and Paraíba; 80% in Ceará; and 62% in Piauí. The others live in simple residences built of adobe (raw adobe), which is one of the oldest techniques used for constructing human habitations, and houses of rammed earth covered by a straw roof. There is a certain prejudice against this construction technique in Brazil due to occurrence of kissing bug that transmits Chagas disease, which houses itself in the holes that appear in the poorly constructed walls of the habitations, particularly in rural areas.

Regarding the water supply, most families do not have access to the public water supply. In Ceará (60%) and Piauí (88%), most of the families access water via artesian wells constructed by the public service. In Bahia, 85.71% have access to water tanks; in Paraíba, 70.59% obtain water from ponds, water holes, dams, small ponds, and water reservoirs, or are supplied by the Brazilian army.

In most of the residences, there was no type of the following basic sanitation: a household sewage disposal system, septic tank and bathrooms. It was even recorded that garbage was improperly left out in the open, buried or burned. Within the rural area of the municipality of Arroio Grande, RS, Brazil, there is no garbage collection in the rural area community, where it is burned on the soil in most cases (45%) by the families, and it can also be placed in a warehouse or cesspool; however, the latter occurs in only a few situations (5%) (Galho et al., 2007).

Among the properties, approximately 92% have electricity in a predominantly single-phase distribution pattern in Bahia, Ceará, and Paraíba, and in a three-phase distribution pattern in Piauí (63%). In only 8% of the residences, there was no access to electricity, and lighting was provided by candles and/or oil lamps.

Families mainly practice agriculture and raise livestock (poultry, goats, and swine), and do so by using primitive and rudimentary methods. This fact can be explained by the lack of technical assistance, financial resources, or even technical knowledge to conduct these activities.

Regarding the lima bean planting time, it is a traditional activity among these families, as it has been practiced there for more than 50 years by 20% of the farmers. The lowest percentage (8%) refers to the newest plantings (1 to 5 years old). According to data of the ABMR&A (2005), 83% of the farmers have dedicated themselves to exploiting the same crop there for more than 10 years.

The study revealed that cultural tradition is one of the predominant factors (92% of the respondents) in choosing the lima bean crop in the municipalities involved in the study followed by the suitability of the crop and soil type (70%), favourable climatic conditions (68%), and regional vocation (60%). The family traditions are very important factors in choosing the common bean crop in Poço Verde, Sergipe State, Brazil, according to the response of 74.19% of the respondents followed by the suitability of the region (22.58%) and lack of opportunities (3.23%) (Lima and Silva Júnior, 2009).

FINAL CONSIDERATIONS

The lima bean farmers are mostly male individuals, married, with an age range between 21 and 75 years, illiterate, with two to three dependents per residence, with farming areas of up to 1 ha, in a leasing system. Additionally, 70% receive public policy contributions. Cultural tradition is the most predominant factor in choosing the lima bean crop, in which the channel most often used for selling the product is the middleman, and most of the production is sold in each given municipality.

The following are the main agricultural production systems observed: intercropped rice, corn, cassava and lima bean, or rice, corn and lima bean, using rudimentary technology, with seeds obtained from the same field and irregular variation in spacing and intercropped arrangements. The production process usually comprises simple techniques, without any type of technological innovation, via techniques passed on from generation to generation.

The lima bean crop is grown at the onset of the rainy period using the "slash and burn" technique and intercropping system. Planting is restricted to the space occupied by trees remaining from the preparation of a clearing. Traditional cultivars are predominantly used; these have an indeterminate growth habit as well as climbing ability. Regarding harvesting, it is mainly performed in July and September, extending until November for the late cultivars.

References

Associação Brasileira de Marketing Rural e Agronegócio - ABMR&A, (2005). *Perfil Comportamental e Hábitos de Mídia do Produtor Rural Brasileiro.*

Arco-Verde, M. F., Mourão, Jr. M., Lopes, C. E. V. & Freitas, F. N. (2004). Diagnóstico sócio-econômico em áreas de pequenos produtores rurais do estado de Roraima, Brasil. *Revista Acadêmica. Ciências Agrárias e Ambientais*, *2*, 61-64.

Azevedo, J. N., Franco, L. J. D. & Araújo, R. O. C. (2003). Composição química de sete variedades de feijão lima. Teresina, PI: *Embrapa Meio-Norte* [*Embrapa Mid-North*], pp 4. (Embrapa Meio-Norte. Comunicado Técnico, 152.).

Baudoin, J. P. (1988). Genetic resources, domestication and evolution of lima bean, Phaseolus lunatus. In: Gepts, P. (ed). *Genetic resources of Phaseolus bean.* Holland, Kluwer Academic Publishers.

Beyra, A. & Artiles, G. R. (2004). Revisión taxonômica de los gêneros *Phaseolus* y *Vigna* (Leguminosae – Papilionoideae) en Cuba. *Anales Del Jardín Botánico de Madrid*, *61*, 135-154.

Bressani, R. & Elias, L. G. Nutritional value of legume crops for humans and animals. In: Summerfield, R.J; Bunting, A.H. (1980). [Ed]. Advances in legume science. Kew: *Royal Botanic Gardens*, 135-155.

Duke, J. A. (1983). Handbook of legumes of world economic importance. 2nd ed. N.York, *Plenum Press.*

Filgueira, F. A. R. (2000). *Novo manual de olericultura: agrotecnologia moderna na produção e comercialização de hortaliças.*, Viçosa, pp 402.

Frazão et al. (2004). Crescimento da lima em resposta a níveis de fósforo na presença e ausência de calagem em Latossolo Amarelo. In: Reunião Brasileira de Fertilidade do Solo e Nutrição de Plantas, 26., 2004, Lajes(SC). *Anais.* Lajes (SC): Sociedade Brasileira de Ciência do Solo.

Fukuda et al. (2001). Diagnóstico rápido participativo dos sistemas de produção da mandioca nos Municípios de Santa Rita, Chapadinha e São Luís, Estado do Maranhão, e curso sobre cultivo e processamento da mandioca. Cruz das Almas, BA: Embrapa Mandioca e Fruticultura /CINPRA São Luís,. pp 25.

Galho et al. (2007). Educação ambiental: O lixo em zona rural do município de Arroio Grande-RS. In: XVI Congresso de Iniciação Científica: Pesquisa e responsabilidade ambiental, Pelotas,. pp 5.

Galvão et al., (2005). Redes solidárias de comercialização e compra de produtos nordestinos no Distrito Federal. *Espaço e Geografia* (UnB), Brasília, *7*, 77-99.

Gândara, F. C. Produção de biomassa e balanço energético em agroecossistemas de produção de hortaliças, no Distrito Federal. Brasília, UnB, 1998. pp 70. (Master's Thesis).

Gerhardt et al. A. (1999). Diagnóstico socioeconômico e ambiental do Município de Maquiné/RS: Perspectivas para um Desenvolvimento Rural Sustentável. In*:* I Salão de Extensão da Universidade Federal do Rio Grande do Sul, Porto Alegre.

Gomes, R. L. F., Lopes, A. C. & de A. (2006). *A cultura da lima*. Comunicado Técnico [Technical Communication]. Teresina: CCA/UFPI, n. 07, pp 1-2.

Instituto Brasileiro de Geografia e Estatística [Brazilian Institute of Geography and Statistics], *IBGE*. Diretoria de Pesquisas, Coordenação de Agropecuária, Produção Agrícola Municipal: mesorregiões, microrregiões e os municípios. Rio de Janeiro, 2014.

Lima, A. A. F. & Silva Júnior, C. D. (2009). Diagnóstico sócio-econômico da produção de feijão no município de Poço Verde - SE. *Revista da Fapese*, *5*, 47-56.

Lymman, J. M. (1983). Adaptation studies on lima bean accessions in Colombia. *Journal of the American Society for Horticultural Science*, Alexandria, *108*, 369-373.

Mariano, J. L. & Neder, H. D. (2004). Renda e Pobreza entre Famílias no meio Rural do Nordeste. In: *IX Encontro Nacional de Economia Política*, Uberlândia. Anais do IX Encontro Nacional de Economia Política.

Mendes, R. H. (2005). Caracterização do perfil dos proprirtários rurais, da região do planalto serrano catarinense. Dissertação apresentada ao Programa de Pós Graduação em Engenharia Florestal da Universidade Federal do Paraná.

Oliveira., et al. (2002). Divergência genética entre cultivares de caupi. *Pesquisa Agropecuária. Brasileira Brasília* 38, 605-611.

Oliveira., et al. (2004). Produção de feijão-lima em função do uso de doses de fósforo em um Neossolo Regolítico. *Horticultura Brasileira* 22, 543-546.

Oliveira Junior., et al. (2007). Sistema Agrícola Consorciado para Agricultura Familiar no Norte do Maranhão. In: *VII Congresso Brasileiro de Sistemas de Produção*, Fortaleza. Agricultura Familiar, Políticas Públicas e Inclusão Social. Fortaleza: SBSP.

Ologhobo, A. D. & Fetuga, B. L. (1983). Compositional differences in some limabean (*Phaseolus lunatus*) varieties. *Food Chemistry*, *10*, 297–307.

Pessoa, C. C. & Souza, M. de; Schuch, I. (2006). Agricultura urbana e segurança alimentar: estudo no município de Santa Maria – RS. Segurança *Alimentar e Nutricional*, *13*, 23-37.

PRRA – *Plano de Reforma Agrária do Estado do Piauí* – 2003/2010. (2005). INCRA, Teresina, pp 58.

Rachie, K. O., Song, L. & Lyman, J. (1980). Lima bean (*Phaseolus lunatus*) and potential in the tropics. In: Summerfield, R. J.; Bunting, A. H. (Eds.). *Advances in legume science.* Kew: Royal Botanic Garden, 375-381.

Ramos, G. M. (2001). Modelo de desenvolvimento sustentável para o Meio-Norte do Brasil: sistema regeneração da agricultura familiar. Teresina: Embrapa Meio-Norte, 73. Il. p. (Embrapa Meio-Norte. Circular Técnica, 31).

Ramos, M. N., Moreira, T. M. & Santos, C. A. (2003). Referências para uma política nacional de educação do campo: caderno de subsídios. Brasília: Secretaria de Educação Média e Tecnológica, Grupo Permanente de Trabalho dc Educação do Campo, pp 48.

Sampaio, F. & Flores, M. (2002). Comércio ético e solidário e a agricultura familiar brasileira: um outro mercado é possível? In: França, C.L. (Org.) Comércio ético e solidário. São Paulo: Fundação Friedrich Ebert / ILDES, 13-30.

Santos., et al. (2002). Produtividade e morfologia de vagens e sementes de variedade de lima no estado da Paraíba [Productivity and morphology of pods and seeds of a lima bean variety in Paraíba State, Brazil]. *Pesquisa Agropecuária Brasileir*, *37*, 1407-1412.

Santos, J. O. (2008). Divergência genética em feijão-lima (*Phaseolus lunatus*) [Genetic divergence in lima bean (*Phaseolus lunatus*)], UFPI, Teresina. pp 97. (Master's thesis).

Silva, E. T. da & Martins, D. E. C. (2004). Perfil dos produtores de produtos hidropônicos do município de Colombo – região metropolitana de Curitiba-Pr. *Revista Acadêmica: ciências agrárias e ambientais* 2, 49-59.

Silva, F. T. C., Barbato, C. E. & Mundim, A. G. (2005). Sensoramento de umidade e temperatura de grãos em sistema armazenados. In: I C*ongresso Interno de Iniciação Científica - PIBIC/CNPq/PIBIC*, 2005, *Cárceres. Caderno de Resumos*, *1*, 177-177.

Silva., et al. (2004). Biodiversidade da Caatinga: áreas e ações prioritárias para a conservação. Brasília, DF: MMA: UFPE, DF: MMA:UFPE:p. Il.

Sindicato Nacional Das Indústrias De Defensivos Agrícolas (2004). SINDAG. Relatório. Brasília.

Soares., et al. (2009). Diagnóstico da cultura do feijão-lima em 11 municípios do Estado do Piauí. *Anais do 5º Congresso Brasileiro de Melhoramento de Plantas*, Guarapari-ES, 1 CD -ROM.

Sousa, I. S. F. de (Org.). (2006). Agricultura Familiar na Dinâmica da Pesquisa Agropecuária. 1ª. ed. Brasília: Embrapa Informação Tecnológica, 1, pp 434.

Vieira, C. (1978). Cultura do feijão. Viçosa: Universidade Federal de Viçosa, , pp 146.

Vieira, D. D. (2007). Identidade rural e agricultura familiar: O pronaf B sob a ótica dos beneficiados. Master's Thesis in Social Sciences: Universidade Federal do Rio Grande do Norte.

Vieira, R.F. A. (1992). cultura do feijão-lima. *Informe Agropecuário* 16, 30-37.

Vieira, R. F. & Araújo, G. A. A. (1992). Comparações agronômicas de feijões dos gêneros Vigna e Phaseolus com o feijão-comum (*Phaseolus Vulgaris* L.) *Pesquisa Agropecuária Brasileira, 27*, 841-850.

Vieira, R. F., Vieira, C. & Vieira, R. F. (2001). *Leguminosas graníferas*. Viçosa: Universidade Federal de Viçosa, pp 206.

EDITORS' CONTACT INFORMATION

Dr. Ademir Sergio Ferreira de Araujo, PhD*
Dr. Angela Celis de Almeida Lopes, PhD
and
Dr. Regina Lucia Ferreira Gomes, PhD

Department of Agricultural Engineering and Soil Science
Agricultural Science Center, Federal University of Piauí
Teresina, PI, Brazil
*Email: asfaruaj@yahoo.com.br

INDEX

A

B

C

D

E

F

G

H

I

O

P

Q

R

S

Y

Z